SpringerBriefs in Molecular Science

Series Editor

Sanjay K. Sharma

For further volumes:
http://www.springer.com/series/8898

Guangming Liu · Guangzhao Zhang

QCM-D Studies on Polymer Behavior at Interfaces

 Springer

Guangming Liu
Department of Chemical Physics
University of Science and Technology
  of China
Hefei
People's Republic of China

Guangzhao Zhang
Faculty of Materials Science and
  Engineering
South China University of Technology
Guangzhou
People's Republic of China

ISSN 2191-5407
ISBN 978-3-642-39789-9
DOI 10.1007/978-3-642-39790-5
Springer Heidelberg New York Dordrecht London

ISSN 2191-5415 (electronic)
ISBN 978-3-642-39790-5 (eBook)

Library of Congress Control Number: 2013945155

© The Author(s) 2013

This work is subject to copyright. All rights are reserved by the Publisher, whether the whole or part of the material is concerned, specifically the rights of translation, reprinting, reuse of illustrations, recitation, broadcasting, reproduction on microfilms or in any other physical way, and transmission or information storage and retrieval, electronic adaptation, computer software, or by similar or dissimilar methodology now known or hereafter developed. Exempted from this legal reservation are brief excerpts in connection with reviews or scholarly analysis or material supplied specifically for the purpose of being entered and executed on a computer system, for exclusive use by the purchaser of the work. Duplication of this publication or parts thereof is permitted only under the provisions of the Copyright Law of the Publisher's location, in its current version, and permission for use must always be obtained from Springer. Permissions for use may be obtained through RightsLink at the Copyright Clearance Center. Violations are liable to prosecution under the respective Copyright Law.
The use of general descriptive names, registered names, trademarks, service marks, etc. in this publication does not imply, even in the absence of a specific statement, that such names are exempt from the relevant protective laws and regulations and therefore free for general use.
While the advice and information in this book are believed to be true and accurate at the date of publication, neither the authors nor the editors nor the publisher can accept any legal responsibility for any errors or omissions that may be made. The publisher makes no warranty, express or implied, with respect to the material contained herein.

Printed on acid-free paper

Springer is part of Springer Science+Business Media (www.springer.com)

Preface

For the solid/liquid interfaces modified with polymers, the interfacial properties would be significantly influenced by the polymer behavior. Understanding the interfacial polymer behavior is a prerequisite to control the interfacial properties and to prepare well-designed polymeric interfacial materials. Actually, the characterization and analysis of polymer behavior at interfaces still remain a great challenge, particularly for the dynamic behavior. In the past 10 years, quartz crystal microbalance with dissipation (QCM-D) has been successfully applied to study the polymer behavior at various solid/liquid interfaces. In comparison with the conventional QCM which can merely provide the change in frequency, QCM-D gives the information about not only the variations in mass, thickness and rigidity of the polymer layer but also the viscoelastic properties of the polymer layer. Therefore, it is anticipated that QCM-D would provide a clearer picture on the molecular interactions of the macromolecular systems at the solid/liquid interfaces.

This book is intended to give an overview of our recent studies on polymer behavior at the solid/liquid interfaces by use of QCM-D. It starts with a brief introduction of the basic principles of QCM-D to give readers a straightforward impression on what is QCM-D and what can be obtained from QCM-D measurements. In the following chapters, we will show that how QCM-D can be used to investigate the conformational change of grafted polymer chains, the grafting kinetics of polymer chains, the growth mechanism of polyelectrolyte multilayers, and the interactions between polymers and phospholipid membranes. The physical meaning of the shifts in frequency and dissipation and the advantages of QCM-D technique will be presented. We hope this book will be helpful to the readers to understand how to use QCM-D to study the polymer behavior at the solid/liquid interfaces. We also hope it can serve as a reference material for academic and industrial researchers working in the fields of polymers and interfaces.

Guangming Liu
Guangzhao Zhang

Contents

1 **Basic Principles of QCM-D** 1
   References 7

2 **Conformational Change of Grafted Polymer Chains** 9
   2.1 Introduction 9
   2.2 Temperature-induced Conformational Change of Grafted
       PNIPAM Chains with a Low Grafting Density 10
   2.3 Temperature-induced Conformational Change of Grafted
       PNIPAM Chains with a High Grafting Density 13
   2.4 Solvency-induced Conformational Change
       of PNIPAM Brushes 15
   2.5 pH-induced Conformational Change of Grafted
       Polyelectrolytes 18
   2.6 Salt Concentration and Type-induced Conformational
       Change of Grafted Polyelectrolytes 23
   2.7 Conclusion 28
   References 29

3 **Grafting Kinetics of Polymer Chains** 33
   3.1 Introduction 33
   3.2 Pancake-to-Brush Transition 34
   3.3 Mushroom-to-Brush Transition 38
   3.4 Conclusion 43
   References 43

4 **Growth Mechanism of Polyelectrolyte Multilayers** 45
   4.1 Introduction 45
   4.2 Roles of Chain Interpenetration and Conformation
       in the Growth of PEMs 47
   4.3 Specific Ion Effect on the Growth of PEMs 54
   4.4 Effects of Chain Rigidity and Architecture
       on the Growth of PEMs 61
   4.5 Conclusion 66
   References 66

5	**Interactions between Polymers and Phospholipid Membranes**	71
	5.1 Introduction	71
	5.2 Role of Hydrophobic Interaction in the Adsorption of PEG on Lipid Membrane Surface	72
	5.3 Effect of Length of Hydrocarbon End Group on the Adsorption of PEG on Lipid Membrane Surface	77
	5.4 Conclusion	80
	References	80

Chapter 1
Basic Principles of QCM-D

Abstract The solution of wave equation relates the eigenfrequency to the thickness of crystal when the acoustic wave propagates in a circular AT-cut quartz crystal, which makes the resonator as a quantitatively ultrasensitive mass sensor possible. When a RF voltage is applied across the electrodes near the resonant frequency, the quartz crystal will be excited to oscillate in the thickness shear mode at its fundamental resonant frequency. For a rigidly adsorbed layer which is evenly distributed and much thinner than the crystal, the added mass on the resonator surface is proportional to the frequency shift (Δf), i.e., they are related by the Sauerbrey equation. On the other hand, the energy dissipation (D) during the oscillation of resonator is an indication of the rigidity of the adsorbed layer. The frequency and dissipation are measured by fitting the oscillation decay of the freely oscillating resonator. In a Newtonian liquid, Δf and ΔD are related not only to the inherent properties of the resonator but also to the solvent viscosity and density. The theoretical representations based on the Voigt model can be used for the viscoelastic film in the liquid medium where the Sauerbrey equation may not be valid. Therefore, QCM-D not only gives the changes in mass and rigidity of the adsorbed layer, but also can provide the information on the viscoelastic properties of the adsorbed layer such as hydrodynamic thickness and shear modulus.

Keywords AT-cut quartz crystal · Sauerbrey equation · Thickness shear mode · Oscillation · Eigenfrequency · Dissipation factor · Viscoelasticity · Voigt model

In 1880, Pierre Curie and his elder brother Jacques Curie found that crystals of Rochelle salt could generate electric potential between opposing surfaces when the crystals were compressed in certain directions (i.e., piezoelectricity) [1]. Two years later, they confirmed that the reverse effect could also occur when the crystals were subjected to an electric field. However, the phenomenon of piezoelectricity and its converse piezoelectric effect did not receive much attention until the World War I when it was demonstrated that quartz crystals could be used as transducers and receivers of ultrasound in water to detect the submarine. In 1921, Cady made the first quartz crystal resonator based on the X-cut crystals [2]. But, the X-cut crystals exhibited very high temperature sensitivity, so that they could

G. Liu and G. Zhang, *QCM-D Studies on Polymer Behavior at Interfaces*,
SpringerBriefs in Molecular Science, DOI: 10.1007/978-3-642-39790-5_1,
© The Author(s) 2013

merely be applied in the fields where the variation of temperature is little importance. The first AT-cut quartz crystal was introduced in 1934, which had nearly zero frequency drift with temperature around room temperature [3]. The advantage of the AT-cut quartz crystal renders this particular cut the most suitable for the mass-detection sensor [4]. In 1959, the linear relationship between the deposited mass and the frequency response was established by Sauerbrey, which formed the fundamental basis of the quartz crystal microbalance (QCM) methodology [5]. Nonetheless, QCM was just used as a mass detector in vacuum or air until the beginning of 1980s when scientists realized that a quartz crystal can be excited with a stable oscillation in a viscous liquid medium [6, 7]. Afterwards, the applications of QCM were extended to many research areas including biology, chemistry, physics, medicine, polymer science, and environmental science [8–10].

Generally, the cut angle of quartz crystal determines the mode of induced mechanical vibration of resonator. Resonators based on the AT-cut quartz crystal with an angle of 35.25° to the optical z-axis would operate in a thickness shear mode (TSM) (Fig. 1.1) [4]. Clearly, the shear wave is a transverse wave, that is, it oscillates in the horizontal direction (x-axis) but propagates in the vertical direction (y-axis). When acoustic waves propagate through a one-dimensional medium, the wave function (ψ) can be described by [11]:

$$\frac{\partial^2 \psi}{\partial x^2} - \frac{1}{v}\frac{\partial^2 \psi}{\partial t^2} = 0 \tag{1.1}$$

where v is the wave speed which depends on the elastic and inertial properties of the medium, x is the position at which the wave function is being described, t is the time. The eigenfrequency (f_n) can be obtained by solving the above equation:

$$f_n = \frac{nv}{2L} \tag{1.2}$$

where n is the overtone number and L is the length of the one-dimensional medium.

Likewise, the wave equation for the propagation of acoustic waves in a three-dimensional medium (e.g., quartz crystal) can be described by [11]:

$$\left(\frac{\partial^2 \psi}{\partial x^2} + \frac{\partial^2 \psi}{\partial y^2} + \frac{\partial^2 \psi}{\partial z^2}\right) - \frac{1}{v}\frac{\partial^2 \psi}{\partial t^2} = 0 \tag{1.3}$$

The corresponding eigenfrequency (f_{nmk}) for the quartz crystal can be calculated by solving the wave equation with appropriate boundary conditions [4]:

$$f_{nmk} = \frac{v}{2}\sqrt{\frac{n^2}{h_q^2} + \frac{m^2}{l_q^2} + \frac{k^2}{w_q^2}} \tag{1.4}$$

where h_q, l_q, and w_q are the thickness, length, and width of the quartz crystal, respectively, and $n, m, k = 1, 3, 5, \ldots$. Obviously, the eigenfrequency of the quartz crystal is determined by its size. Similarly, the eigenfrequency for the acoustic

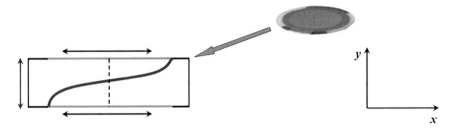

Fig. 1.1 Schematic illustration of the thickness-shear-mode resonator, where the shear wave oscillates in the *horizontal direction* but propagates in the *vertical direction*

waves traveling in a circular AT-cut quartz crystal with a radius of r_q can be obtained by solving the wave equation using the cylindrical coordinates [4]:

$$f_{nmk} = \frac{v}{2\pi}\sqrt{\frac{n^2\pi^2}{h_q^2} + \frac{\chi_{mk}^2}{r_q^2}} \quad (1.5)$$

where $n = 1, 3, 5, \ldots$, $m = 0, 1, 2, 3, \ldots$, $k = 1, 2, 3, \ldots$, and χ_{mk} is the kth root of Bessel function of order m. If $r_q \gg h_q$, the propagation of the shear wave can be treated one-dimensionally [4]:

$$f_n \approx \frac{v}{2\pi}\sqrt{\frac{n^2\pi^2}{h_q^2}} = \frac{nv}{2h_q} = \frac{n}{2h_q}\sqrt{\frac{\mu_q}{\rho_q}} = nf_0 \quad (1.6)$$

where f_0 is the fundamental resonant frequency, μ_q is the shear modulus of quartz, ρ_q is the density of quartz, and $v = (\mu_q/\rho_q)^{1/2}$. It can be seen from Eq. (1.6) that the fundamental resonant frequency is mainly influenced by the thickness of the quartz crystal since other parameters are constant in this equation, which makes the AT-cut quartz crystal as a quantitatively ultrasensitive mass sensor possible. The QCM sensor generally has a sandwich structure, namely, the quartz crystal is placed between a pair of gold electrodes (Fig. 1.1). When a radio frequency (RF) voltage is applied across the electrodes near the resonant frequency, the quartz crystal will be excited to oscillate in the thickness shear mode at its fundamental resonant frequency. Assuming that the addition of a thin layer to the electrodes will induce a change in the crystal thickness from h_q to h_q', then the frequency will change from f_n to f_n':

$$f_n' = \frac{nv}{2h_q'} \quad (1.7)$$

Therefore, the frequency shift (Δf) caused by the deposition of the additional layer can be expressed as:

$$\Delta f = f_n' - f_n = -\frac{nv(h_q' - h_q)}{2h_q h_q'} = -\frac{nv\Delta h_q}{2h_q h_q'} \quad (1.8)$$

By the combination of Eqs. (1.6) and (1.8), we can get:

$$\frac{\Delta f}{f_n} = -\frac{\Delta h_q}{h'_q} \quad (1.9)$$

If $\Delta h_q \ll h_q$, Eq. (1.9) can be written as:

$$\frac{\Delta f}{f_n} \approx -\frac{\Delta h_q}{h_q} = -\frac{\Delta h_q \rho_q A_q}{h_q \rho_q A_q} = -\frac{\Delta M_q}{h_q \rho_q A_q} \quad (1.10)$$

Equation (1.10) describes the relationship between the frequency shift and the mass change of the quartz crystal (ΔM_q). A_q is the area of the electrode on the crystal surface. For a rigidly adsorbed layer which is evenly distributed and much thinner than the crystal, ΔM_q is approximately equal to the mass change induced by the adsorbed layer (ΔM_f) [5, 6], so that Eq. (1.10) can be described as:

$$\frac{\Delta f}{f_n} = \frac{\Delta f}{nf_0} = -\frac{\Delta M_f}{h_q \rho_q A_q} = -\frac{\Delta m_f}{h_q \rho_q} \quad (1.11)$$

where Δm_f is the mass change of the adsorbed layer per unit area (i.e., areal mass density, $\Delta m_f = \Delta M_f/A_q$). Then, the so-called Sauerbrey equation can be derived from Eq. (1.11) [5]:

$$\Delta m_f = -\frac{\rho_q h_q}{f_0}\frac{\Delta f}{n} = -C\frac{\Delta f}{n} \quad (1.12)$$

This equation relates the mass change of the adsorbed layer to the frequency shift of the quartz crystal, which forms the fundamental basis of the highly sensitive QCM mass detection technique. Here, C is the mass sensitivity constant of ~ 17.7 ng cm^{-2} Hz^{-1} for a 5 MHz quartz resonator [12].

However, the Sauerbrey equation can be merely used to estimate the mass change of a rigidly adsorbed layer on the resonator surface in air or vacuum. When a viscoelastic film is deposited on the resonator surface in liquid medium, the oscillation of resonator would be damped by the adsorbed layer. If the damping in the deposited film becomes sufficiently large, the linear relationship between Δf and Δm_f is no longer valid [13]. Therefore, this requires defining another parameter to characterize the viscoelastic properties of the adsorbed layer. The energy dissipation during the oscillation of resonator can be described with the dissipation factor (D) [14]:

$$D = \frac{1}{Q} = \frac{E_d}{2\pi E_s} \quad (1.13)$$

where Q is the quality factor of the crystal, E_d is the energy dissipated during one oscillation, and E_s is the energy stored in the oscillating system. A larger value of D reflects the formation of a softer and more swollen layer, whereas a smaller D indicates a relatively rigid and dense layer adsorbed on the resonator surface [15].

1 Basic Principles of QCM-D

The change in D can be obtained by measuring the impedance spectroscopy [9] or by fitting the oscillation decay [14]. In the former method, a broader resonance peak is indicative of a larger dissipation factor. In the latter case, the measurement of D is based on the fact that the amplitude of oscillation (A) or the output voltage over the crystal decays as an exponentially damped sinusoid when the driving power of the piezoelectric oscillator is switched off at $t = 0$ (Fig. 1.2) [16]:

$$A(t) = A_0 e^{-t/\tau} \sin(2\pi f t + \varphi) + C \tag{1.14}$$

where τ is the decay time constant, $A(t)$ is the amplitude of oscillation at time of t, A_0 is the amplitude at $t = 0$, C is a constant. By fitting the oscillation decay of the freely oscillating resonator, one can get the frequency (f). Note that the output f is the difference between the resonant frequency (f_0) of the crystal and the constant reference frequency (f_r), i.e., $f = f_0 - f_r$ [13]. The value of D can be obtained by the following relation [14]:

$$D = \frac{1}{\pi f \tau} \tag{1.15}$$

In this book, all the studies on the polymer behavior at the solid/liquid interfaces are conducted on a quartz crystal microbalance with dissipation (QCM-D) from Q-sense AB [14]. The measurements of the shifts in frequency (Δf) and dissipation (ΔD) are based on the second method. Thus, we can simultaneously obtain a series of changes of Δf and ΔD via fitting the oscillation decay by

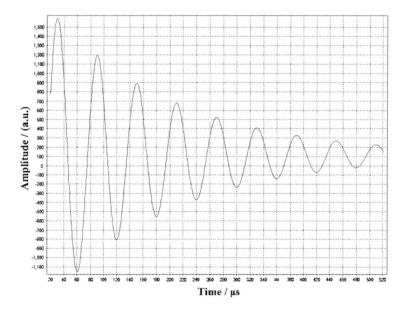

Fig. 1.2 The exponential decay of the amplitude of oscillation of a 5 MHz, AT-cut quartz crystal in water at 25 °C

switching the driving voltage on and off periodically. When the quartz resonator is immersed in a Newtonian liquid, the frequency response of the quartz resonator can be quantitatively described by the Kanazawa-Gordon equation [17]:

$$\Delta f = -n^{\frac{1}{2}} f_0^{\frac{3}{2}} (\eta_l \rho_l / \pi \mu_q \rho_q)^{\frac{1}{2}} \quad (1.16)$$

where ρ_l and η_l are the density and viscosity of the liquid medium, respectively. On the other hand, the change in dissipation factor in a Newtonian liquid can be expressed as [18]:

$$\Delta D = 2(f_0/n)^{\frac{1}{2}} (\eta_l \rho_l / \pi \mu_q \rho_q)^{\frac{1}{2}} \quad (1.17)$$

Equations (1.16) and (1.17) show that Δf and ΔD are related not only to the inherent properties of the quartz crystal but also to the solvent viscosity and density. Therefore, the shifts in Δf and ΔD induced by the polymer behavior at the resonator surface can be extracted by taking the background response of the blank resonator as a Ref. [19].

Since the Sauerbrey equation is not valid for the viscoelastic layer, theoretical representations based on the Voigt model can be applied in such situations [20]. Assuming that the adsorbed layer surrounded by a semi-infinite Newtonian fluid under no-slip conditions is homogenous with a uniform thickness, the complex shear modulus (G) of the adsorbed layer can be described by:

$$G = G' + iG'' = \mu_f + i2\pi f \eta_f = \mu_f (1 + i2\pi f \tau_f) \quad (1.18)$$

where G' is the storage modulus, G'' is the loss modulus, μ_f is the elastic shear modulus, η_f is the shear viscosity, τ_f is the characteristic relaxation time of the film which equals to the ratio of η_f to μ_f. Then, Δf and ΔD can be given by:

$$\Delta f = \mathrm{Im}\left(\frac{\beta}{2\pi \rho_q l_q}\right) \quad (1.19)$$

and

$$\Delta D = -\mathrm{Re}\left(\frac{\beta}{\pi f \rho_q l_q}\right) \quad (1.20)$$

where

$$\beta = \xi_1 \frac{2\pi f \eta_f - i\mu_f}{2\pi f} \frac{1 - \alpha \exp(2\xi_1 h_f)}{1 + \alpha \exp(2\xi_1 h_f)}, \quad \alpha = \frac{\frac{\xi_1}{\xi_2} \frac{2\pi f \eta_f - i\mu_f}{2\pi f \eta_l} + 1}{\frac{\xi_1}{\xi_2} \frac{2\pi f \eta_f - i\mu_f}{2\pi f \eta_l} - 1},$$

$$\xi_1 = \sqrt{-\frac{(2\pi f)^2 \rho_f}{\mu_f + i2\pi \eta_f}}, \quad \xi_2 = \sqrt{i \frac{2\pi f \rho_f}{\eta_l}},$$

and where ρ_f and h_f are density and thickness of the adsorbed layer, respectively. Based on the theoretical representations mentioned above, the hydrodynamic thickness (h_f), the shear viscosity (η_f), and the shear modulus (μ_f) can be obtained by fitting the changes of Δf and ΔD at different overtones using the Voigt model with a Q-tools software from Q-sense AB [20].

For solid surfaces coated with a polymer layer immersed in liquid media, the interfacial properties would be significantly influenced by the polymer behavior at the interfaces. So far, the characterization and analysis of polymer behavior at the solid/liquid interfaces still remain a great challenge, particularly for the dynamic polymer behavior. From the brief introduction of QCM-D technique above, one can recognize that QCM-D not only gives us the information about the changes in mass, thickness, and rigidity of the adsorbed layer, but tells us how the viscoelastic properties of the adsorbed layer vary on the resonator surface. In the following chapters, we will intend to give an overview of our recent studies on the polymer behavior at the solid/liquid interfaces by use of QCM-D. We will show that how QCM-D can be used to investigate the conformational change of grafted polymers, the grafting kinetics of polymer chains, the growth mechanism of polyelectrolyte multilayers, and the interactions between polymers and lipid membranes. In all the studies, the physical meaning of the changes in frequency and dissipation will be clearly interpreted and the advantages of QCM-D technique will also be demonstrated. We hope the readers can more profoundly understand the QCM-D technique after reading this book.

References

1. Curie J, Curie P (1880) Développement par pression de l'électricité polaire dans les cristaux hémièdres à faces inclinées. Compt Rendus 91:294–295
2. Cady WG (1921) The piezoelectric resonator. Phys Rev A 17:531–533
3. Lack FR, Willard GW, Fair IE (1934) Some improvements in quartz crystal circuit elements. Bell Syst Tech J 13:453–463
4. Bottom VE (1982) Introduction to quartz crystal unit design. Van Nostrand Reinhold Co., New York
5. Sauerbrey G (1959) Verwendung von svhwingquarzen zur wägung dünner schichten und zur mikrowägung. Z Phys 155:206–222
6. Lu C, Czanderna AW (1984) Applications of piezoelectric quartz crystal microbalances. Elsevier, Amsterdam
7. Nomura T, Okuhara M (1982) Frequency shifts of piezoelectric quartz crystals immersed in organic liquids. Anal Chim Acta 142:281–284
8. Buttry DA, Ward MD (1992) Measurement of interfacial processes at electrode surfaces with the electrochemical quartz crystal microbalance. Chem Rev 92:1355–1379
9. Janshoff A, Galla HJ, Steinem C (2000) Piezoelectric mass-sensing devices as biosensors—an alternative to optical biosensors? Angew Chem Int Ed 39:4004–4032
10. Liu GM, Zhang GZ (2008) Applications of quartz crystal microbalance in polymer science. Chin Polym Bull 8:174–188
11. Feynman R (1969) Lectures in physics. Addison Publishing Company, Addison

12. Höök F, Kasemo B (2001) Variations in coupled water, viscoelastic properties, and film thickness of a Mefp-1 protein film during adsorption and cross-linking: a quartz crystal microbalance with dissipation monitoring, ellipsometry, and surface plasmon resonance study. Anal Chem 73:5796–5804
13. Steinem C, Janshoff A (2007) Piezoelectric sensors. Springer, Berlin
14. Rodahl M, Höök F, Krozer A, Brzezinski P, Kasemo B (1995) Quartz crystal microbalance setup for frequency and Q-factor measurements in gaseous and liquid environments. Rev Sci Instrum 66:3924–3930
15. Höök F, Rodahl M, Brzezinski P, Kasemo B (1998) Energy dissipation kinetics for protein and antibody-antigen adsorption under shear oscillation on a quartz crystal microbalance. Langmuir 14:729–734
16. Smith KL (1986) Q-Ratio. Electron Wireless W 92:51–53
17. Kanazawa KK, Gordon JG (1985) Frequency of a quartz microbalance in contact with liquid. Anal Chem 57:1770–1771
18. Rodahl M, Kasemo B (1996) On the measurement of thin liquid overlayers with the quartz-crystal microbalance. Sensor Actuators A-Phys 54:448–456
19. Zhang GZ, Wu C (2009) Quartz crystal microbalance studies on conformational change of polymer chains at interface. Macromol Rapid Commun 30:328–335
20. Voinova MV, Rodahl M, Jonson M, Kasemo B (1999) Viscoelastic acoustic response of layered polymer films at fluid-solid interfaces: continuum mechanics approach. Phys Scr 59:391–396

Chapter 2
Conformational Change of Grafted Polymer Chains

Abstract Based on the "grafting from" and "grafting to" methods, polymer chains are grafted onto the resonator surfaces. QCM-D is used to investigate the conformational change of grafted chains induced by the variation of external conditions. For the grafted poly(N-isopropylacrylamide) (PNIPAM) chains, the QCM-D studies show that the conformational change of grafted PNIPAM chains induced by the variations of temperature and solvent composition is fundamentally different from that for the free PNIPAM chains in solution and the grafting density plays an important role in the conformational change. For the grafted polyelectrolytes, the chemical oscillation induced periodic collapse and swelling of poly (acrylic acid) brushes and the pH-induced folding of DNA with different grafting densities are discussed in detail with the QCM-D results. The influences of salt concentration and salt type on the conformational change of grafted polyelectrolytes are also discussed in this chapter. The studies demonstrate that QCM-D can provide not only the changes in mass and rigidity of the grafted polymer chains, but also the changes in hydrodynamic thickness, shear viscosity, and shear modulus of the grafted polymer layer, which would give a clear picture on the conformational change of the grafted polymer chains.

Keywords Conformational change · Polymer brushes · Polyelectrolyte · Hydration · Electrostatic interaction · Folding · Counterion condensation · Charge reversion

2.1 Introduction

It is well-known that the interfacial properties are significantly influenced by the conformation of polymer chains grafted at interfaces [1–4]. However, the in situ real-time characterization of the conformational change of grafted chains still remains a challenge [5, 6]. Generally, polymer chains at the solid/liquid interface would exhibit rich conformations depending on the polymer structure, grafting density, solvent quality, and polymer segment–surface interaction [7–11]. At a low grafting density where the distance between the grafted chains is larger than the size of the chains, the grafted chains usually exhibit a pancake-like conformation when polymer segments attractively interact with the surface [12]. In contrast, if there are no obviously attractive segment–surface interactions, the chains would have a mushroom structure [12]. At a high grafting density where the distance between the grafted chains is less than the chain size, the chains will form brushes [12]. For charged chains, the conformation is also influenced by the electrostatic interactions between the grafted polyelectrolyte chains [13].

Due to the stimuli-responsive properties of polymers, the conformation of grafted polymer chains will be strongly dependent on the external conditions such as temperature, pH, salt concentration, and salt type [14]. Such stimuli-responsive properties have important implications and applications in the field of smart interfacial materials [15]. In comparison with free chains in solution, polymer chains grafted on a solid surface are expected to exhibit distinct conformational change upon the variation of external environments [16]. In this chapter, we will show that how the QCM-D can be used to investigate the conformational change of grafted neutral and charged polymer chains induced by varying temperature, solvent composition, pH, salt concentration, and salt type [17–24].

2.2 Temperature-induced Conformational Change of Grafted PNIPAM Chains with a Low Grafting Density

Poly (N-isopropylacrylamide) (PNIPAM) is a well-known thermally sensitive polymer and has a lower critical solution temperature (LCST) at ~ 32 °C in water [25]. That is, individual PNIPAM chains adopt a random coil conformation at low temperatures but collapse into a globule when the solution temperature is higher than LCST. In contrast with a linear PNIPAM chain free in solution exhibiting a discontinuous coil-to-globule transition [26], the grafted PNIPAM chains may exhibit different conformational changes due to the constraint of the chains on surface [27]. Some studies demonstrated that grafted PNIPAM chains has a sharp transition near the LCST [28, 29], but other investigations indicated a gradual collapse of the grafted PNIPAM chains over the LCST [30], which agreed with the theoretical prediction [9, 27]. To clarify the question, PNIPAM chains are grafted onto the QCM resonator surface to investigate the collapse and swelling of the grafted chains induced by the variation of temperature.

Actually, PNIPAM chains can be anchored on the solid surface by either "grafting to" or "grafting from" method [31]. The former generally would lead to the grafted polymer chains with a low grafting density, whereas the latter usually results in the grafted polymer chains with a high grafting density [17, 19].

2.2 Temperature-induced Conformational Change of Grafted PNIPAM Chains

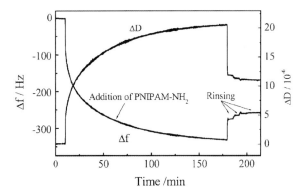

Fig. 2.1 Time dependence of changes in frequency (Δf) and dissipation (ΔD) during the grafting of PNIPAM chains onto the gold-coated resonator surface at T of 20 °C, where the overtone number n = 3. Reprinted with the permission from Ref. [17]. Copyright 2004 American Chemical Society

PNIPAM chains are first grafted onto the QCM resonator surface with a low grafting density according to the "grafting to" procedure [17]. Figure 2.1 shows a real-time measurement of Δf and ΔD for the addition of linear NH_2-PNIPAM to the QCM chamber where the resonator surface is immobilized with the groups of -$S(CH_2)_{12}OCH_2COON(CO)_2(CH_2)_2$. The decrease of Δf as well as the increase of ΔD with time indicates the grafting of PNIPAM chains onto the resonator surface. The physically adsorbed chains can be removed by rinsing with water.

Figure 2.2 shows the temperature dependence of $-\Delta f$. The shift in Δf is indicative of the mass change of the grafted polymer layer [32]. $-\Delta f$ gradually decreases with the increase of temperature during the heating process, indicating that the mass of the polymer layer on the resonator surface decreases with temperature. Here, the mass detected by QCM includes the mass of the grafted PNIPAM chains and the coupled water molecules. Since PNIPAM chains are grafted on the surface, the decrease in $-\Delta f$ implies the dehydration of the polymer chains, that is, some bounded water molecules leave the grafted PNIPAM layer. During the cooling process, $-\Delta f$ increases with the decreasing temperature,

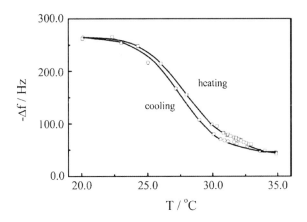

Fig. 2.2 Temperature dependence of frequency shift ($-\Delta f$) of linear PNIPAM chains grafted on the gold-coated resonator surface, where the overtone number n = 3. Reprinted with the permission from Ref. [17]. Copyright 2004 American Chemical Society

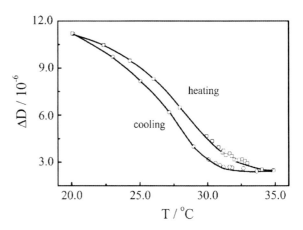

Fig. 2.3 Temperature dependence of dissipation shift $(-\Delta D)$ of linear PNIPAM chains grafted on the gold-coated resonator surface, where the overtone number n = 3. Reprinted with the permission from Ref. [17]. Copyright 2004 American Chemical Society

indicating that the PNIPAM chains rehydrate again as temperature decreases. Eventually, $-\Delta f$ is back to the original point, suggesting that PNIPAM chains can resume complete hydration at the low temperature.

The temperature dependence of ΔD is shown in Fig. 2.3. It is known that a dense and rigid layer has a small dissipation, whereas a layer with a loose and flexible structure exhibits a large dissipation [33]. It can be seen that ΔD decreases with temperature during the heating process, indicating that PNIPAM chains shrink and collapse into a denser structure. During the cooling process, ΔD increases with the decreasing temperature, indicating the swelling of PNIPAM layer. Additionally, a hysteresis can be observed in the heating-and-cooling cycle in either Fig. 2.2 or 2.3. This is because PNIPAM segments form additional hydrogen bonds at the collapsed state, which cannot be completely removed at the temperature near LCST [34].

On the other hand, the relation of ΔD versus Δf can describe the cooperativity between the conformational change and the hydration of the grafted polymer chains because Δf mainly arises from the hydration/dehydration of polymer chains while ΔD is due to the swelling/collapse of the polymer layer. The fact that ΔD linearly increases with $-\Delta f$ in Fig. 2.4 indicates that the conformational change involves only one kinetic process, which suggests that the dehydration and collapse occur simultaneously in the heating process and the rehydration is concurrent with the swelling in the cooling process. In other words, they have a strong cooperativity during the heating/cooling process.

Fig. 2.4 The relation between ΔD and $-\Delta f$ for the temperature-induced conformational change of the PNIPAM chains grafted on the gold-coated resonator surface, where the overtone number n = 3. Reproduced from Ref. [6] by permission of John Wiley & Sons Ltd

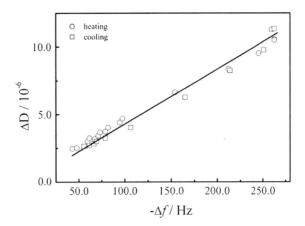

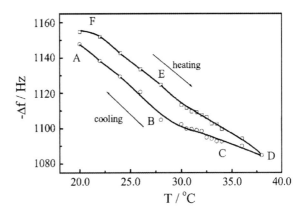

Fig. 2.5 Temperature dependence of frequency shift ($-\Delta f$) of PNIPAM brushes grafted on the SiO$_2$-coated resonator surface, where the overtone number n = 3. Reprinted with the permission from Ref. [19]. Copyright 2005 American Chemical Society

2.3 Temperature-induced Conformational Change of Grafted PNIPAM Chains with a High Grafting Density

To understand the influence of grafting density on the conformational change of grafted polymer chains, PNIPAM chains are grafted onto a SiO$_2$-coated resonator surface via the "grafting from" procedure based on the surface-initiated polymerization method to prepare the PNIPAM brushes with a high grafting density [19]. Figure 2.5 shows the temperature dependence of $-\Delta f$ of PNIPAM brushes in one heating-and-cooling cycle. It can be seen that $-\Delta f$ gradually decreases with the increasing temperature in the heating process over the range of 20–38 °C, which is similar to the observation in Fig. 2.2. At low temperatures, water is a good solvent for PNIPAM, so that PNIPAM chains strongly interact with water molecules. As temperature increases, dehydration occurs and PNIPAM chains gradually collapse, leading to the decrease of $-\Delta f$. In contrast, the collapsed

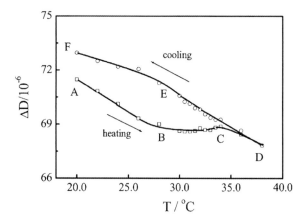

Fig. 2.6 Temperature dependence of dissipation shift (ΔD) of PNIPAM brushes grafted on the SiO$_2$-coated resonator surface, where the overtone number n = 3. Reprinted with the permission from Ref. [19]. Copyright 2005 American Chemical Society

PNIPAM brushes become swollen and rehydrate again with the decreasing temperature in the cooling process, as indicated by the increase of $-\Delta f$. Obviously, the continuous change of frequency in both heating and cooling processes is markedly different from the discontinuous coil-to-globule transition of individual PNIPAM chains free in water [26].

Figure 2.6 shows the temperature dependence of ΔD in one heating and cooling cycle. ΔD decreases with the increasing temperature in the heating process, indicating that PNIPAM brushes gradually collapse into a more compact structure. In the cooling process, ΔD increases with the decreasing temperature, indicating that the collapsed brushes become more swollen and flexible. Also, ΔD in the cooling process is larger than that in the heating process at the same temperature, whereas an opposite trend is observed for $-\Delta f$ in Fig. 2.5. This phenomenon should arise from the formation of tails on the outer layer of the brushes, which have significant effects on the dissipation. Specifically, the brushes begin to swell from their outer layer to inner core in the cooling process, resulting in some

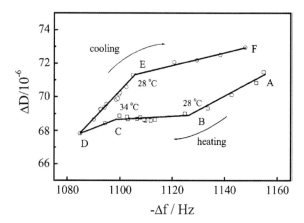

Fig. 2.7 The relation between ΔD and $-\Delta f$ for the PNIPAM brushes grafted on the SiO$_2$-coated resonator surface during the temperature induced conformational change, where the overtone number n = 3. Reprinted with the permission from Ref. [19]. Copyright 2005 American Chemical Society

2.3 Temperature-induced Conformational Change of Grafted PNIPAM Chains

random and flexible tails on the layer surface, thereby giving rise to a larger ΔD than that in the heating process. The collapsed PNIPAM brushes cannot be completely back to the initial swollen state during the experimental time scale, as indicated by the fact that both ΔD and $-\Delta f$ are not back to their original values at 20 °C, which is different from that for the grafted PNIPAM chains with a low grafting density (Figs. 2.2 and 2.3).

In the relation between ΔD and $-\Delta f$ (Fig. 2.7), three kinetic processes can be observed in the heating process. When $T < 28$ °C (A to B), ΔD decreases with the decrease of $-\Delta f$, implying the concurrent shrinking and dehydration of grafted PNIPAM chains. In the range 28 °C $< T <$ 34 °C (B to C), ΔD slightly changes as $-\Delta f$ decreases. Here, PNIPAM brushes are partially collapsed in this range of temperature. The conformational change of the partially collapsed brushes is limited due to the steric barrier, so that ΔD slightly changes. On the other hand, not all the detached water molecules during shrinking at $T < 28$ °C can leave PNIPAM brushes immediately, some of them are trapped in the dense brushes. As temperature increases, the trapped water molecules gradually diffuse out of the brushes, leading to an obvious decrease of $-\Delta f$. Obviously, the cooperativity between the collapse and the dehydration is weak due to the retarded dehydration, which is also responsible for the continuous collapse transition. Further heating would overcome the steric barrier, leading to more collapse and dehydration, as reflected by the decreases in ΔD and $-\Delta f$ at $T > 34$ °C (C to D). In the cooling process, when the temperature is higher than 28 °C (D to E), ΔD rapidly increases with the increasing $-\Delta f$, suggesting that the flexible tails are formed on the outer layer of PNIPAM brushes during the rehydration. When $T < 28$ °C (E to F), ΔD slows down its increase because the hydrated chains in the outer layer tend to stretch and pack more densely. For the same $-\Delta f$, the value of ΔD in the cooling process is always higher than that in the heating process, further indicating that the flexible tails have a pronounced effect on the dissipation. The hysteresis observed in the heating and cooling cycle is attributed to the additional hydrogen bonds formed in the collapsed state. Besides, the nonuniformity and the stretching of the brushes are thought to further enlarge the hysteresis. In fact, the poly (N-isopropylacrylamide-co-sodium acrylate) copolymer brushes also exhibit a similar conformational change when temperature is varied [20].

2.4 Solvency-induced Conformational Change of PNIPAM Brushes

PNIPAM not only has the thermosensitive property, but also exhibits a reentrant behavior or cononsolvency in response to the composition of a mixed solvent consisting of water and a water-miscible organic solvent [35]. That is, PNIPAM is soluble in either water or organic solvent but precipitant in their mixture with a certain composition. Several models have been proposed to explain the reentrant

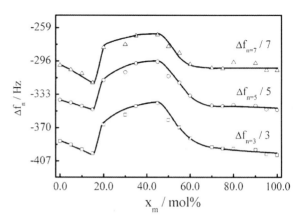

Fig. 2.8 Methanol molar fraction (x_m) dependence of frequency shift (Δf_n) of the PNIPAM brushes at different overtones (n = 3, 5, 7). Reprinted with the permission from Ref. [18]. Copyright 2005 American Chemical Society

behavior, including the perturbation of the water–methanol interaction parameter (χ) by PNIPAM chains [36], the formation of complexes between water and methanol [37], and the formation of competitive hydrogen bonds between PNIPAM chains and solvent molecules [38]. To have a better understanding of the reentrant behavior, QCM-D is employed to investigate the conformational change of PNIPAM brushes on a SiO_2-coated resonator surface induced by varying the solvent composition of water–methanol mixtures [18].

Figure 2.8 shows the methanol molar fraction (x_m) dependence of frequency shift (Δf_n) of PNIPAM brushes at different overtones. For the sake of comparison, the frequency shift measured at each overtone is divided by n. The overtone dependent Δf_n indicates that PNIPAM brushes have a viscoelastic nature. The Δf_n in response to the composition of the water–methanol mixtures at different overtones exhibits the same trend. At $x_m \sim 17$ %, Δf_n sharply increases, indicating the sharp decrease of the number of solvent molecules bound to PNIPAM chains. The gradual increase of Δf_n in the range of x_m between 20 and 45 % reveals a continuous desolvation of the grafted PNIPAM chains. Further increase of the methanol content leads to the resolvation of the grafted PNIPAM chains, as indicated by the sharp decrease of Δf_n at $x_m \sim 50$ % and the following slow decrease at $x_m > 60$ %. Obviously, when the x_m is in the range of 17–50 %, the water–methanol mixtures are poor solvents for PNIPAM. The solvation-to-desolvation-to-resolvation transition reflects the swelling-to-collapse-to-swelling of the grafted PNIPAM layer.

Figure 2.9 shows x_m dependence of ΔD of the PNIPAM brushes. In contrast to Δf, there is a sharp decrease of ΔD at $x_m \sim 17$ %, indicating the collapse of the PNIPAM brushes. The gradual decrease of ΔD in the range of x_m between 20 and 45 % indicates a further collapse of the PNIPAM brushes. The sharp increase at $x_m \sim 50$ % reflects that the collapsed PNIPAM brushes re-swell into a looser layer.

The cooperativity between the collapse/swelling and the desolvation/solvation of the PNIPAM brushes can be viewed by the relation of ΔD versus Δf. Figure 2.10 shows that ΔD linearly decreases as Δf increases, suggesting that the

2.4 Solvency-induced Conformational Change of PNIPAM Brushes

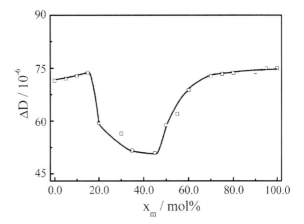

Fig. 2.9 Methanol molar fraction (x_m) dependence of dissipation shift (ΔD) of the PNIPAM brushes, where the overtone number n = 3. Reprinted with the permission from Ref. [18]. Copyright 2005 American Chemical Society

reentrant transition involves only one kinetic process. This fact reveals that the desolvation and collapse occur simultaneously and the solvation is concurrent with the swelling. Namely, the cooperativity is strong in the present system. This fact also indicates that no preferential solvation occurs here, which is quite different from the case of polymer brushes in a mixture of good and poor solvents [39]. The reentrant transition of the PNIPAM brushes in the mixture of water and methanol is probably due to the formation of water/methanol complexes or clusters consisting of a certain number of water and methanol molecules, which is very sensitive to the solvent composition. Such a complexation leads the water–methanol mixture to change suddenly from a good solvent to a poor one for PNIPAM at $x_m \sim 17\%$, and develops into a good solvent again at $x_m \sim 50\%$. In such a solvent mixture, either PNIPAM brushes or individual PNIPAM chains are expected to undergo a sharp reentrant transition.

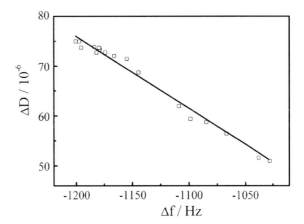

Fig. 2.10 The relation between ΔD and Δf for the solvency induced conformational change of the PNIPAM brushes, where the overtone number n = 3. Reprinted with the permission from Ref. [18]. Copyright 2005 American Chemical Society

2.5 pH-induced Conformational Change of Grafted Polyelectrolytes

Polyelectrolytes are macromolecules carrying ionic groups, which can dissociate in water or other polar solvents forming charged polymer backbones [40]. When polyelectrolyte chains are grafted on a solid surface, the physical properties of the grafted charged polymer chains are different from that of the grafted uncharged polymer chains [13]. In addition to the intra- and intermolecular interactions involved in the neutral polymer system, electrostatic interactions come into play in the determination of the conformation of polyelectrolyte brushes. It is known that the conformation of polyelectrolyte brushes is strongly influenced by salt concentration and salt type [41]. For the weak polyelectrolytes, the conformation is also influenced by the solution pH [42]. Generally, the switch of conformation of polyelectrolyte brushes is stimulated by manually varied external conditions, that is, the stimuli-responsive properties of the charged polymer layer act in an equilibrium state. However, many physiological behaviors such as heartbeat and brainwaves exhibit rhythmical oscillations. Thus, it is also interesting to see how the conformation of polyelectrolyte brushes is changed driven by a chemical oscillation.

Poly (acrylic acid) (PAA) brushes are prepared on a gold-coated resonator surface by using surface-initiated atom transfer radical polymerization method [21]. The QCM-D measurements on the pH-induced oscillation of PAA brushes under a continuous flow of bromate-sulfite-ferrocyanide (BSF) solution are shown in Fig. 2.11 [21]. Figure 2.11a shows the time dependence of pH of BSF solution. Typically, pH is varied between 3.2 and 6.6. Parts b and c of Fig. 2.11 show the time dependence of shifts in Δf and ΔD of PAA brushes under a continuous flow of BSF solution. Clearly, both Δf and ΔD exhibit obvious oscillations. As pH increases from 3.2 to 6.6, Δf rapidly decreases. Then, Δf sharply increases as pH varies from 6.6 to 3.2. In contrast, the response of dissipation oscillates with pH in an opposite trend. PAA is a weak polyelectrolyte with $pK_a \sim 4.5$ [43]. Therefore, as pH increases from 3.2 to 6.6, more carboxyl groups are ionized, so more water molecules are coupled with PAA chains, leading Δf to decrease. In contrast, as pH decreases from 6.6 to 3.2, Δf increases due to the dehydration of PAA chains. In addition, the slight overtone dependence at pH 3.2 implies PAA brushes are less viscoelastic, whereas the stronger dependence at pH 6.6 indicates that the brushes are more viscoelastic. This can be viewed more clearly from the change in ΔD. As pH increases from 3.2 to 6.6, ΔD increases, indicating that more energy of resonator oscillation is damped by PAA brushes, namely, the brushes become more viscoelastic, whereas the decrease of ΔD with pH varied from 6.6 to 3.2 indicates the collapse of PAA brushes.

Figure 2.12 shows the changes in hydrodynamic thickness (Δt_f), shear viscosity (η_f), and elastic shear modulus (μ_f) of PAA brushes estimated by using the Voigt model. Due to the stretching and collapse of PAA brushes, Δt_f oscillates between

2.5 pH-induced Conformational Change of Grafted Polyelectrolytes

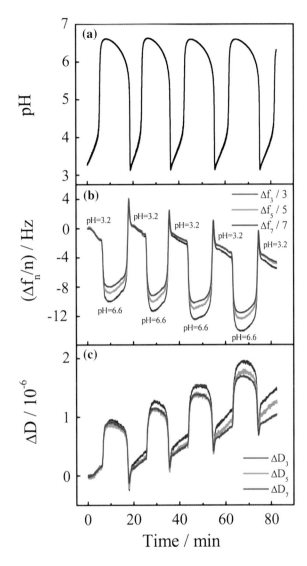

Fig. 2.11 a Oscillation of pH of the BSF solution, where [KBrO$_3$] = 0.065 M, [Na$_2$SO$_3$] = 0.075 M, K$_4$[Fe(CN)$_6$] = 0.02 M, [H$_2$SO$_4$] = 0.01 M. **b** Time dependence of shift in frequency (Δf_n) of PAA brushes under a continuous flow of BSF solution, where the overtone number $n = 3$, 5, 7. **c** Time dependence of shift in dissipation (ΔD) of PAA brushes under a continuous flow of BSF solution, where the overtone number $n = 3, 5, 7$. Reprinted with the permission from Ref. [21]. Copyright 2008 American Chemical Society

pH 3.2 and 6.6 with amplitude of ~2 nm. At the same time, η_f and μ_f also exhibit oscillations with the change of pH. As pH increases from 3.2 to 6.6, η_f increases from ~1.0 × 10^{-3} to ~2.7 × 10^{-3} Pa·s due to the hydration of PAA chains. On the other hand, the increase of electrostatic repulsions between PAA chains with pH varied from 3.2 to 6.6 leads PAA brushes to adopt a weakly compressible state, so that μ_f increases from ~1.0 × 10^4 to ~2.8 × 10^5 Pa. In contrast, η_f and μ_f decrease with the decreasing pH.

In comparison with the synthetic polyelectrolytes, studies on the conformational change of biopolyelectrolytes (e.g., DNA) have more important implications

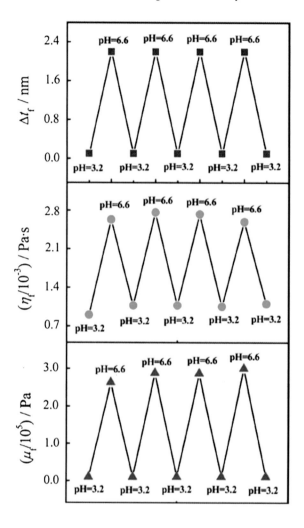

Fig. 2.12 Changes in hydrodynamic thickness (Δt_f), shear viscosity (η_f), and elastic shear modulus (μ_f) of PAA brushes under the oscillation of pH. Reprinted with the permission from Ref. [21]. Copyright 2008 American Chemical Society

for the biological activity of cellular biomacromolecules [44]. DNA with cytidine-enriched sequences can fold into an i-motif via intercalated C–C$^+$ base pairing in an acidic environment, whereas it unfolds into a random coil in a basic environment [45]. Consequently, the folding and unfolding of DNA can be simply modulated by tuning solution pH. The DNA chains are grafted onto a gold-coated resonator surface at pH 4.5 and 8.5 based on the thiol-gold reaction, which, respectively, gives rise to DNA chains with low and high grafting densities [22]. The grafted DNA chains are immersed in a solution at pH 8.5 first so that they are able to completely unfold. Subsequently, a solution with pH of 4.5 is introduced to the QCM chamber to replace the high pH solution, which will lead to the folding of grafted DNA chains. Figure 2.13 shows Δf quickly increases and then gradually levels off when pH is changed from 8.5 to 4.5. This is an indication of dehydration

2.5 pH-induced Conformational Change of Grafted Polyelectrolytes

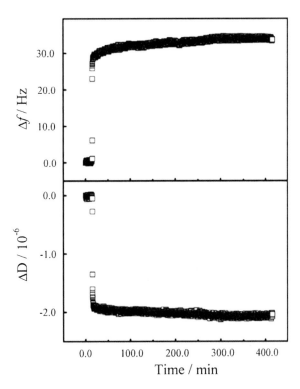

Fig. 2.13 Shifts in frequency (Δf) and dissipation (ΔD) for the folding of grafted DNA chains with a low grafting density induced by changing pH, where the overtone number n = 3. Reprinted with the permission from Ref. [22]. Copyright 2010 American Chemical Society

during the folding of DNA chains. It is known that the protonated cytidine groups at pH 4.5 would form intra-/interchain hydrogen bonds due to C–C$^+$ pairing [46]. The interactions lead to the release of some coupled water molecules from DNA chains, resulting in an increase in Δf. On the other hand, the quick decrease in ΔD indicates the collapse of DNA chains during the folding.

Figure 2.14 shows the shifts in Δf and ΔD for the folding of DNA chains with a high grafting density. Obviously, there are several stages during the folding of grafted DNA chains. Δf increases and ΔD decreases quickly in the initial stage because the pH-induced folding leads to the dehydration and collapse of the DNA chains. Then, Δf and ΔD only slightly change, which might be due to the slight rearrangement of grafted DNA chains. Afterwards, Δf increases and ΔD decreases gradually with time. As stated above, because of the high grafting density, the DNA chains restrict each other to fold into the i-motif structure. The DNA chains have to gradually fold as they need to rearrange and shape themselves. After a long time, Δf and ΔD level off, indicating the completion of the folding.

The folding of DNA chains can be viewed in terms of ΔD versus $-\Delta f$ relation (Fig. 2.15). The folding of DNA chains with a low grafting density only involves one kinetic process. This is because the space around DNA chains is enough for their full folding. However, two kinetic processes are observed in the folding of

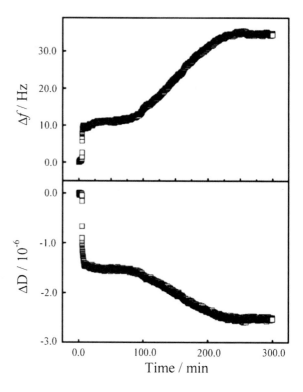

Fig. 2.14 Shifts in frequency (Δf) and dissipation (ΔD) for the folding of grafted DNA chains with a high grafting density induced by changing solution pH, where the overtone number n = 3. Reprinted with the permission from Ref. [22]. Copyright 2010 American Chemical Society

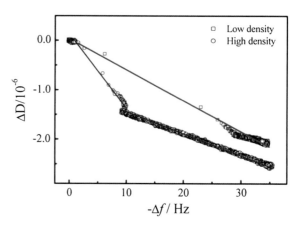

Fig. 2.15 The relation between ΔD and $-\Delta f$ for the folding of DNA chains at different grafting densities, where the overtone number n = 3. Reprinted with the permission from Ref. [22]. Copyright 2010 American Chemical Society

DNA with a high grafting density. As discussed above, the DNA chains are expected to partially fold in the first process. In the second process, the DNA chains have to rearrange themselves to have more space for further folding.

2.6 Salt Concentration and Type-induced Conformational Change of Grafted Polyelectrolytes

One interesting and unclear question regarding polyelectrolytes is the reentrant condensation of polyelectrolyte chains upon addition of multivalent salt [47]. Namely, polyelectrolyte chains form a precipitate as the concentration of multivalent salt increases, and the precipitate will redissolve into solution upon the further addition of multivalent salt. The precipitation of polyelectrolyte chains in the presence of multivalent salts limits the insight into the mechanism for the reentrant behavior at the molecular level by using some techniques (e.g., laser light scattering). However, the macroscopic phase separation in the presence of multivalent counterions which occurs for the free chains in solution can be avoided in the QCM-D measurements for the grafted polyelectrolyte chains, so that we can look insight the microscopical mechanism of the reentrant behavior [23].

Figure 2.16 shows the ionic strength (I) dependence of Δf for the resonator grafted with sodium poly(styrene sulfonate) (PSS) chains in different divalent salt solutions. The data obtained in $CaCl_2$ solution are used as an example to discuss. At $I < 1.5$ M, Δf decreases as I increases from 3.0×10^{-4} to 9.0×10^{-2} M (A to C), indicating the trapping of water molecules by an inhomogeneous layer formed by the grafted PSS chains. From C to D, I increases from 9.0×10^{-2} to 1.5 M, the charges on PSS chains are almost completely neutralized by the counterions. Thus, the chains are dehydrated and collapsed to form a dense and homogenous layer and the trapped water molecules are released out, as indicated by the increase of Δf. At $I > 1.5$ M, Δf decreases again as I increases (D to E), implying that the mass associated with the PSS brushes increases. This is an indication of the rehydration and reexpansion of the grafted PSS chains. For the divalent salts, when the PSS chains collapse into the homogeneous layer, the adsorbed counterions are expected to form a strongly correlated liquid layer at the polyelectrolyte surface due to the strong counterion–counterion correlation [48]. Such a liquid layer has a larger

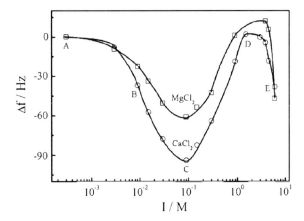

Fig. 2.16 The ionic strength (I) dependence of frequency shift (Δf) for the resonator grafted with PSS chains in different divalent salt solutions, where the overtone number n = 3. Reproduced from Ref. [23] by permission of The Royal Society of Chemistry

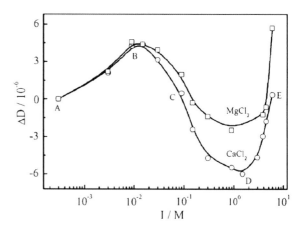

Fig. 2.17 The ionic strength (I) dependence of dissipation shift (ΔD) for the resonator grafted with PSS chains in different divalent salt solutions, where the overtone number n = 3. Reproduced from Ref. [23] by permission of The Royal Society of Chemistry

dielectric constant (ε_1) than that of water (ε_2), therefore the counterions near the polyelectrolyte surface cause polarization of the surface under the liquid layer and produce "image" charges with opposite sign on the polyelectrolyte surface based on the equation $q' = q(\varepsilon_2 - \varepsilon_1)/(\varepsilon_2 + \varepsilon_1)$, where q and q' are the original charge and image charge, respectively [48]. At the same time, the repulsion between the incoming counterions and the adsorbed counterions would create some correlation holes [48]. As I increases (D to E), the attractions between the incoming counterions and the image charges give rise to further counterion condensation onto the PSS chains through the correlation holes, which causes the charge inversion of the PSS chains. In other words, the PSS chains now become positively charged. Such a recharging leads to the rehydration of the chains, and the electrostatic repulsion between the PSS chains results in the reexpansion of PSS brushes. This is why Δf decreases from D to E. Additionally, from A to C, a larger change in Δf is observed for $CaCl_2$ than that for $MgCl_2$. This can be explained by the formation of relatively strong ion pairs between Ca^{2+} and the weakly hydrated alkyl sulfonate groups since the strength of hydration of Ca^{2+} is weaker than Mg^{2+} [49].

In Fig. 2.17, the change in ΔD further demonstrates the conformational change of grafted PSS chains in $CaCl_2$ and $MgCl_2$ solutions. Specifically, ΔD increases with the increasing I from 3.0×10^{-4} to 9.0×10^{-3} M (A to B), indicating the formation of a loose and inhomogeneous layer. As I increases from 9.0×10^{-3} to 9.0×10^{-2} M (B to C), the decrease in ΔD indicates that the layer becomes denser. Further increasing I from 9.0×10^{-2} to 1.5 M (C to D) leads the grafted PSS chains to form a even more homogeneous and denser layer, as indicated by the further decrease of ΔD from C to D. At $I > 1.5$ M, the increase of I (D to E) leads ΔD to increase again, indicating the reexpansion of the PSS layer due to the charge inversion.

Figure 2.18 shows the $-\Delta f$ dependence of ΔD as a function of ionic strength for the resonator grafted with PSS chains in $CaCl_2$ solution. ΔD increases with $-\Delta f$ (A to B), implying that the highly extended PSS chains partly shrink into a loose and inhomogeneous structure. From B to C, ΔD decreases but $-\Delta f$ increases, indicating

2.6 Salt Concentration and Type-induced Conformational Change

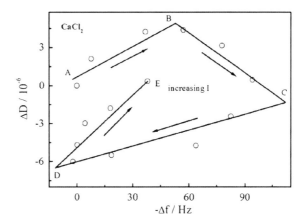

Fig. 2.18 The frequency shift $(-\Delta f)$ dependence of dissipation shift (ΔD) as a function of ionic strength (I) for the resonator grafted with PSS chains in the CaCl$_2$ solutions, where the overtone number n = 3. Reproduced from Ref. [23] by permission of The Royal Society of Chemistry

that the inhomogeneous layer forms a relatively dense structure but the water molecules are still trapped in such a structure. Further increasing I (C to D) leads the PSS chains to form a denser and more homogeneous structure, as reflected by the fact that both $-\Delta f$ and ΔD decrease as I increases. The increase in either $-\Delta f$ or ΔD in the last regime (D to E) indicates the rehydration and reexpansion of the grafted PSS chains due to the charge inversion. The conformational change of grafted PSS chains in the presence of trivalent salts has similar results [23].

From discussion above, counterion condensation plays a crucial role in determining polyelectrolyte conformational change. Actually, the extent of counterion condensation is influenced not only by the salt concentration, but also by the charge density of polyelectrolyte chains [50]. For the charged chains, the counterions condense around the oppositely charged groups on the polyelectrolyte chains due to the electrostatic attraction, screening the electrostatic repulsion between the identically charged groups, leading to the collapse of the chains. When the charge density decreases to zero, the polyelectrolytes become uncharged chains. The nonelectrostatic force such as van der Waals force between the ions and nonpolar moiety of polyelectrolyte chains may have effects on the conformational change [51]. In contrast with the electrostatic ion-polar group interaction that gives rise to the "salting-in" effect by charging the group, nonelectrostatic ion-nonpolar surface interaction can result in a "salting-out" effect by dehydrating the nonpolar moiety or a "salting-in" effect by binding onto the nonpolar moiety with respect to different ions [52]. Poly[(2-dimethylamino)ethyl methacrylate] (PDEM) is a weak polyelectrolyte whose charge density can be tuned by pH, and the chains are completely charged, partially charged, and uncharged at pH 4, 7, and 10, respectively [53]. PDEM chains are grafted onto a gold-coated resonator surface by the "grafting to" procedure, and then the conformational change of the grafted PDEM chains can be investigated using QCM-D in salt solutions at different pH [24].

Figure 2.19 shows I dependence of Δf and ΔD for the resonator grafted with PDEM chains immersed in Na$_2$SO$_4$ and NaClO$_3$ solutions at pH 4. The increase in

Fig. 2.19 Ionic strength (*I*) dependence of frequency shift (Δ*f*) and dissipation shift (Δ*D*) for the resonator grafted with PDEM chains immersed in Na$_2$SO$_4$ and NaClO$_3$ solutions at pH 4, where the overtone number n = 3. Reprinted with the permission from Ref. [24]. Copyright 2011 American Chemical Society

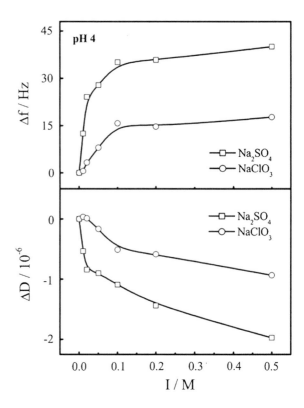

Δ*f* and decrease in Δ*D* with the increasing *I* indicate the dehydration of the grafted chains and the release of water molecules trapped between the chains upon the counterion condensation, which is accompanied by the chain collapse. In the range of *I* < 0.1 M, the grafted chains exhibit a more rapid collapse than that in the higher ionic strength regime, as reflected by the relatively rapid decrease in Δ*D* in the lower ionic strength regime. On the other hand, for the same ionic strength, SO$_4^{2-}$ can more effectively induce the chain collapse than ClO$_3^-$ does. This is understandable because the divalent counterion can bind with two charged groups at most. As a result, the grafted chains can be cross-linked by SO$_4^{2-}$, leading to more effective dehydration and chain collapse. Obviously, the conformational change of grafted PDEM chains at pH 4 is dominated by the counterion condensation.

Figure 2.20 shows *I* dependence of Δ*f* and Δ*D* for the resonator grafted with PDEM chains immersed in Na$_2$SO$_4$ and NaClO$_3$ solutions at pH 7. When pH is increased from 4 to 7, PDEM chains become partially charged and the electrostatic repulsion is weakened, so that the chains adopt partially collapsed conformation. For both salts, Δ*f* gradually increases as *I* increases, indicating the dehydration of the chains and the release of water molecules from the PDEM layer. This leads the polyelectrolyte chains to further collapse, as indicated by the decrease of

2.6 Salt Concentration and Type-induced Conformational Change

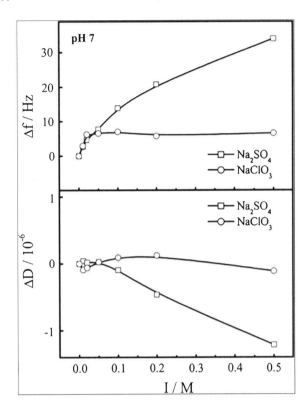

Fig. 2.20 Ionic strength (I) dependence of frequency shift (Δf) and dissipation shift (ΔD) for the resonator grafted with PDEM chains immersed in Na$_2$SO$_4$ and NaClO$_3$ solutions at pH 7, where the overtone number n = 3. Reprinted with the permission from Ref. [24]. Copyright 2011 American Chemical Society

ΔD. Again, SO$_4{}^{2-}$ causes larger shifts in Δf and ΔD than ClO$_3{}^-$ does. However, the difference is smaller than that at pH 4. This is because PDEM chains at pH 7 are already partially collapsed and the addition of salts can only induce limited conformational change of the chains. Clearly, the conformational change of grafted PDEM chains at pH 7 is also dominated by the counterion condensation.

Figure 2.21 shows I dependence of Δf and ΔD for the resonator grafted with PDEM chains immersed in Na$_2$SO$_4$ and NaClO$_3$ solutions at pH 10. It can be seen that Δf exhibits a slight increase with I in the range of $I < \sim 0.05$ M, followed by a gradual decrease with the further increase of I for both SO$_4{}^{2-}$ and ClO$_3{}^-$. Likewise, as I increases, ΔD first decreases and then gradually increases. Considering that the nonpolar hydrophobic moiety of polymer chains is hydrated by the surrounding water molecules in aqueous solution, the approach of SO$_4{}^{2-}$ or ClO$_3{}^-$ to the nonpolar moiety may dehydrate the surface (salting-out effect) as the adsorption of such two anions would increase the surface tension of the hydrophobic surface [52]. This is why we observe an increase in Δf with the increasing I in the low ionic strength regime. The chain dehydration would strengthen the hydrophobic force between the chain segments, causing a further chain collapse, as indicated by the fact that ΔD decreases with I in the low ionic strength regime. When the ionic strength is above ~ 0.05 M, Δf gradually decreases with I,

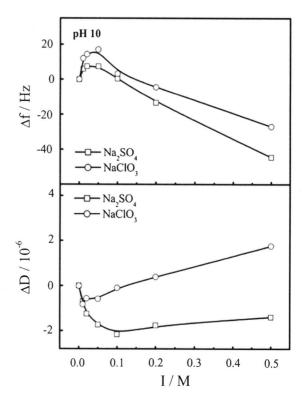

Fig. 2.21 Ionic strength (*I*) dependence of frequency shift (Δf) and dissipation shift (ΔD) for the resonator grafted with PDEM chains immersed in Na_2SO_4 and $NaClO_3$ solutions at pH 10, where the overtone number n = 3. Reprinted with the permission from Ref. [24]. Copyright 2011 American Chemical Society

suggesting the rehydration of the chains due to the binding of anions onto the dimethylamino groups of PDEM (salting-in effect) [54]. The gradual increase of ΔD with *I* implies that the bound anions lead to some swelling of the PDEM layer. Besides, Δf decreases and ΔD increases linearly as *I* increases in the range of *I* > 0.05 M. This is because the strength of the salting-in interaction between ions and polar groups depends on ionic strength [55]. Clearly, the conformational change of the grafted PDEM chains at pH 10 is dominated by the anion–nonpolar moiety interaction and the anion–polar group interaction in the low and high ionic strength regimes, respectively.

2.7 Conclusion

In this chapter, QCM-D is used to investigate the conformational change of the grafted polymer chains. Generally, QCM-D can provide not only the information about the mass change of the grafted chains, but also those about the structural change of the grafted layer. More specifically, Δf is related to the solvation/desolvation of the grafted chains and ΔD is correlated with the stretching/collapse

2.7 Conclusion

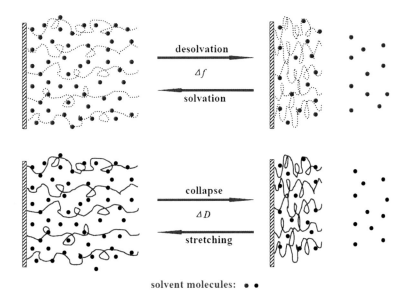

Fig. 2.22 Schematic illustration of the relation between the frequency shift (Δf) and the solvation/desolvation of the grafted polymer chains and the relation between the dissipation shift (ΔD) and the stretching/collapse of the grafted polymer chains

of the grafted chains (Fig. 2.22). The combination of Δf and ΔD can provide more useful information, e.g., the kinetic processes of the conformational change. Based on the Voigt model, QCM-D can also tell how the hydrodynamic thickness, the shear viscosity, and the shear modulus change during the conformational change.

References

1. Bhat RR, Tomlinson MR, Wu T, Genzer J (2006) Surface-grafted polymer gradients: formation, characterization, and applications. Adv Polym Sci 198:51–124
2. Ionov L, Minko S (2012) Mixed polymer brushes with locking switching. Acs Appl Mater Interfaces 4:483–489
3. Moya S, Azzaroni O, Farhan T, Osborne VL, Huck WT (2005) Locking and unlocking of polyelectrolyte brushes: toward the fabrication of chemically controlled nanoactuators. Angew Chem Int Ed 44:4578–4581
4. Neuhaus S, Padeste C, Spencer ND (2011) Versatile wettability gradients prepared by chemical modification of polymer brushes on polymer foils. Langmuir 27:6855–6861
5. Barbey R, Lavanant L, Paripovic D, Schuwer N, Sugnaux C, Tugulu S, Klok HA (2009) Polymer brushes via surface-initiated controlled radical polymerization: Synthesis, characterization, properties, and applications. Chem Rev 109:5437–5527
6. Zhang GZ, Wu C (2009) Quartz crystal microbalance studies on conformational change of polymer chains at interface. Macromol Rapid Commun 30:328–335
7. Alexander S (1977) Adsorption of chain molecules with a polar head a-scaling description. J Phys-Paris 38:983–987

8. de Gennes PG (1980) Conformations of polymers attached to an interface. Macromolecules 13:1069–1075
9. Milner ST (1991) Polymer brushes. Science 251:905–914
10. Halperin A, Tirrell M, Lodge TP (1992) Tethered chains in polymer microstructures. Adv Polym Sci 100:31–71
11. Zhao B, Brittain WJ (2000) Polymer brushes: surface-immobilized macromolecules. Prog Polym Sci 25:677–710
12. Fleer GJ, Cohen Stuart MA, Scheutjens JMHM, Cosgrove T, Vincent B (1993) Polymers at Interfaces. Chapman & Hall, UK
13. Rühe J, Ballauff M, Biesalski M, Dziezok P, Grohn F, Johannsmann D, Houbenov N, Hugenberg N, Konradi R, Minko S, Motornov M, Netz RR, Schmidt M, Seidel C, Stamm M, Stephan T, Usov D, Zhang HN (2004) Polyelectrolyte brushes. Adv Polym Sci 165:79–150
14. Tagliazucchi M, Szleifer I (2012) Stimuli-responsive polymers grafted to nanopores and other nano-curved surfaces: structure, chemical equilibrium and transport. Soft Matter 8:3292–3305
15. Stuart MAC, Huck WTS, Genzer J, Muller M, Ober C, Stamm M, Sukhorukov GB, Szleifer I, Tsukruk VV, Urban M, Winnik F, Zauscher S, Luzinov I, Minko S (2010) Emerging applications of stimuli-responsive polymer materials. Nat Mater 9:101–113
16. Dukes D, Li Y, Lewis S, Benicewicz B, Schadler L, Kumar SK (2010) Conformational transitions of spherical polymer brushes: synthesis, characterization, and theory. Macromolecules 43:1564–1570
17. Zhang GZ (2004) Study on conformation change of thermally sensitive linear grafted poly (N-isopropylacrylamide) chains by quartz crystal microbalance. Macromolecules 37:6553–6557
18. Liu GM, Zhang GZ (2005) Reentrant behavior of poly (N-isopropylacrylamide) brushes in water–methanol mixtures investigated with a quartz crystal microbalance. Langmuir 21:2086–2090
19. Liu GM, Zhang GZ (2005) Collapse and swelling of thermally sensitive poly (N-isopropylacrylamide) brushes monitored with a quartz crystal microbalance. J Phys Chem B 109:743–747
20. Cheng H, Liu GM, Wang CQ, Zhang GZ, Wu C (2006) Collapse and swelling of poly (N-isopropylacrylamide-co-sodium acrylate) copolymer brushes grafted on a flat SiO_2 surface. J Polym Sci Polym Phys 44:770–778
21. Liu GM, Zhang GZ (2008) Periodic swelling and collapse of polyelectrolyte brushes driven by chemical oscillation. J Phys Chem B 112:10137–10141
22. Xia HW, Hou Y, Ngai T, Zhang GZ (2010) pH induced DNA folding at interface. J Phys Chem B 114:775–779
23. Hou Y, Liu GM, Wu Y, Zhang GZ (2011) Reentrant behavior of grafted poly (sodium styrenesulfonate) chains investigated with a quartz crystal microbalance. Phys Chem Chem Phys 13:2880–2886
24. Wang XW, Liu GM, Zhang GZ (2011) Conformational behavior of grafted weak polyelectrolyte chains: effects of counterion condensation and nonelectrostatic anion adsorption. Langmuir 27:9895–9901
25. Schild HG (1992) Poly (N-isopropylacrylamide)-experiment, theory and application. Prog Polym Sci 17:163–249
26. Wu C, Zhou SQ (1995) Laser-light scattering study of the phase-transition of poly (N-isopropylacrylamide) in water. 1 Single-chain. Macromolecules 28:8381–8387
27. Grest GS, Murat M (1994) Monte carlo and molecular dynamics simulations in polymer science. In: Binder K (ed). Clarendon, Oxford
28. Takei YG, Aoki T, Sanui K, Ogata N, Sakurai Y, Okano T (1994) Dynamic contact-angle measurement of temperature-responsive surface-properties for poly (N-Isopropylacrylamide) grafted surfaces. Macromolecules 27:6163–6166
29. Zhang J, Pelton R, Deng YL (1995) Temperature-dependent contact angles of water on poly (N-isopropylacrylamide) gels. Langmuir 11:2301–2302

References

30. Balamurugan S, Mendez S, Balamurugan SS, O'Brien MJ, Lopez GP (2003) Thermal response of poly (N-isopropylacrylamide) brushes probed by surface plasmon resonance. Langmuir 19:2545–2549
31. Minko S (2008) Polymer surfaces and interfaces. In: Stamm M. (ed). Springer, Berlin
32. Sauerbrey G (1959) Verwendung von svhwingquarzen zur wägung dünner schichten und zur mikrowägung. Z Phys 155:206–222
33. Voinova MV, Rodahl M, Jonson M, Kasemo B (1999) Viscoelastic acoustic response of layered polymer films at fluid-solid interfaces: continuum mechanics approach. Phys Scr 59:391–396
34. Wu C, Zhou SQ (1995) Thermodynamically stable globule state of a single poly (N-isopropylacrylamide) chain in water. Macromolecules 28:5388–5390
35. Winnik FM, Ringsdorf H, Venzmer J (1990) Methanol water as a co-nonsolvent system for poly (N-isopropylacrylamide). Macromolecules 23:2415–2416
36. Amiya T, Hirokawa Y, Hirose Y, Li Y, Tanaka T (1987) Reentrant phase-transition of N-isopropylacrylamide gels in mixed-solvents. J Chem Phys 86:2375–2379
37. Zhang GZ, Wu C (2001) The water/methanol complexation induced reentrant coil-to-globule-to-coil transition of individual homopolymer chains in extremely dilute solution. J Am Chem Soc 123:1376–1380
38. Tanaka F, Koga T, Winnik FM (2008) Temperature-responsive polymers in mixed solvents: competitive hydrogen bonds cause cononsolvency. Phys Rev Lett 101:028302
39. Auroy P, Auvray L (1992) Collapse-stretching transition for polymer brushes-preferential solvation. Macromolecules 25:4134–4141
40. Forster S, Schmidt M (1995) Polyelectrolytes in solution. Adv Polym Sci 120:51–133
41. Zhou F, Hu HY, Yu B, Osborne VL, Huck WTS, Liu WM (2007) Probing the responsive behavior of polyelectrolyte brushes using electrochemical impedance spectroscopy. Anal Chem 79:176–182
42. Schuwer N, Klok HA (2011) Tuning the pH sensitivity of poly (methacrylic acid) brushes. Langmuir 27:4789–4796
43. Gebhardt JE, Fuerstenau DW (1983) Adsorption of polyacrylic-acid at oxide water interfaces. Colloid Surf 7:221–231
44. Williamson DH, Denny PW, Moore PW, Sato S, McCready S, Wilson RJM (2001) The in vivo conformation of the plastid DNA of *Toxoplasma gondii*: implications for replication. J Mol Biol 306:159–168
45. Gueron M, Leroy JL (2000) The i-motif in nucleic acids. Curr Opin Struc Biol 10:326–331
46. Simmel FC, Dittmer WU (2005) DNA nanodevices. Small 1:284–299
47. Hsiao PY, Luijten E (2006) Salt-induced collapse and reexpansion of highly charged flexible polyelectrolytes. Phys Rev Lett 97:148301
48. Grosberg AY, Nguyen TT, Shklovskii BI (2002) Colloquium: the physics of charge inversion in chemical and biological systems. Rev Mod Phys 74:329–345
49. Collins KD (2004) Ions from the Hofmeister series and osmolytes: effects on proteins in solution and in the crystallization process. Methods 34:300–311
50. Manning GS (2007) Counterion condensation on charged spheres, cylinders, and planes. J Phys Chem B 111:8554–8559
51. Bostrom M, Williams DRM, Ninham BW (2002) The influence of ionic dispersion potentials on counterion condensation on polyelectrolytes. J Phys Chem B 106:7908–7912
52. Satoh M, Kawashima T, Komiyama J (1991) Competitive counterion binding and dehydration of polyelectrolytes in aqueous-solutions. Polymer 32:892–896
53. An SW, Thomas RK (1997) Determination of surface pKa by the combination of neutron reflection and surface tension measurements. Langmuir 13:6881–6883
54. Maison W, Kennedy RJ, Kemp DS (2001) Chaotropic anions strongly stabilize short, N-capped uncharged peptide helicies: a new look at the perchlorate effect. Angew Chem Int Ed 40:3819–3821
55. Baldwin RL (1996) How Hofmeister ion interactions affect protein stability. Biophys J 71:2056–2063

Chapter 3
Grafting Kinetics of Polymer Chains

Abstract The chemical grafting of thiol-terminated poly(N-isopropylacrylamide) (HS-PNIPAM) chains to a gold-coated resonator surface from an aqueous solution is investigated by using QCM-D in real time. The frequency and dissipation responses reveal that the HS-PNIPAM chains exhibit three-regime kinetics of the grafting. In regimes I and II, the PNIPAM chains form pancake and mushroom structures, respectively. In regime III, the chains form brushes. The grafting of thiol-terminated poly[(2-dimethylamino)ethyl methacrylate] (HS-PDEM) chains to the gold-coated resonator surface is also investigated by using QCM-D in real time. The frequency and dissipation responses demonstrate that the three-regime kinetics can also be observed during the grafting of PDEM chains. The chains are quickly grafted in regime I forming a random mushroom. In regime II, the grafted chains undergo a rearrangement and form an ordered mushroom structure. Further increasing the grafting density leads the chains to form brushes in regime III. For either HS-PNIPAM or HS-PDEM, the mushroom-to-brush transition occurs from regime II to III.

Keywords Pancake · Mushroom · Brush · Electrostatic interaction · Conformational change · Grafting density · Chain elasticity · Adsorption

3.1 Introduction

Polymer chains grafted on a surface will exhibit rich conformations depending on the grafting density and the polymer segment-surface interactions [1]. At a low grafting density, because the distance between grafting sites is larger than the size of the chains, the grafted polymer chains do not overlap. If the polymer segments have an attractive interaction with the surface, the polymer chains will exhibit a "pancake-like" conformation. In contrast, if the segment-surface interaction is nonattractive, a "mushroom" structure can be observed. At a high grafting density, where the distance between grafting sites is smaller than the size of the chains, as the result of the balance between segment–segment repulsion and elasticity of the chains, the grafted chains are stretched away from the surface to form brushes.

Most of the studies have been focused on the properties of the grafted chains in the static equilibrium [2]. Only a few studies have been conducted on the kinetics of grafting [3–8]. Using scaling arguments, Alexander predicted that the polymer chains are adsorbed on the surface, forming a two-dimensional pancake structure at a low grafting density [9]. As the grafting density increases, some of the trains, tails, and loops desorb from the surface, but they do not overlap each other, and the chains form an unstretched three-dimensional mushroom structure. When the grafting density is high enough, a first-order pancake-to-brush transition occurs, where the chains repel each other and stretch out [9]. Ligoure also predicted a first-order phase transition, but it was expected to occur at a higher surface coverage than that predicted by Alexander [10]. On the other hand, Penn et al. experimentally investigated the kinetics of the mushroom-to-brush transition and three-regime kinetics was observed in their experiments [6–8]. However, the results are inconsistent with the theoretical predictions where only two distinct regimes are expected, namely, the chains form mushroom and brush structures in the first and second regimes, respectively [4].

3.2 Pancake-to-Brush Transition

The narrowly distributed thiol-terminated PNIPAM (HS-PNIPAM) is prepared by using reversible addition-fragmentation chain transfer (RAFT) polymerization and the subsequent hydrolysis [11, 12], so that the chains can be chemically grafted on the gold-coated resonator surface because of the strong chemical coupling between the SH groups and the gold surface [13]. The PNIPAM segment-surface interaction is examined by measuring the adsorption of dithiobenzoate group terminated PNIPAM (DTB-PNIPAM) because such an interaction has significant effects on the kinetics of grafting. The only difference between DTB-PNIPAM and HS-PNIPAM is in the terminated group. Figure 3.1 shows the addition of DTB-PNIPAM induces large changes in Δf (~ -180 Hz) and ΔD ($\sim 4.8 \times 10^{-6}$) after rinsing. This fact indicates that PNIPAM segments can be absorbed on the gold surface because of the strong segment-surface interactions. Therefore, it is expected that a pancake-to-brush transition can be realized in this system with the increasing grafting density [12].

Figure 3.2 shows the frequency shift of the resonator immersed in an aqueous HS-PNIPAM solution as a function of logarithmic time. The grafting stopped when the gold surface was saturated at ~ 6300 min. Obviously, the grafting exhibits a three-regime-kinetics character. In the initial stage, Δf significantly decreases (regime I), indicating that the chains are quickly grafted on the bare gold surface. The slow decrease of Δf in regime II suggests that the chains are gradually grafted on the surface because the already grafted chains hinder the further grafting. An accelerated grafting occurs in regime III, as reflected by the relatively

3.2 Pancake-to-Brush Transition

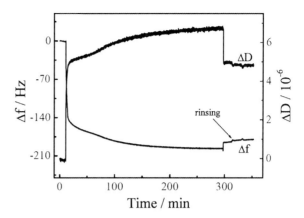

Fig. 3.1 Time dependence of shifts in frequency (Δf) and dissipation (ΔD) of the gold-coated resonator immersed in a DTB-PNIPAM solution, where the overtone number $n = 3$. Reprinted with the permission from Ref. [12]. Copyright 2005 American Chemical Society

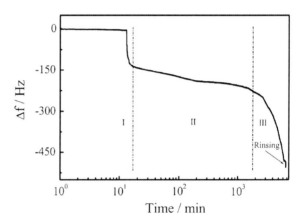

Fig. 3.2 Frequency shift (Δf) of the resonator immersed in a HS-PNIPAM solution as a function of logarithmic time, where the overtone number $n = 3$. Reprinted with the permission from Ref. [12]. Copyright 2005 American Chemical Society

sharp decrease in Δf, which implies that the conformation of the already grafted chains is changed to accommodate the incoming chains.

Figure 3.3 shows the dissipation shift of the resonator immersed in the HS-PNIPAM solution as a function of logarithmic time. The dissipation shift can provide information on the structural change of the polymer layer on the surface. The dissipation of the resonator should increase with the thickness and looseness of the grafted polymer layer. That is, a dense and rigid structure has a small dissipation, while a loose and thick structure leads to a large dissipation. The sharp increase in ΔD in regime I further indicates the quick grafting of the chains. In regime II, the small increase in dissipation indicates a slow grafting due to the steric barrier created by the already grafted chains. The large increase in ΔD in regime III is indicative of the acceleration of the grafting and the increase of thickness of the polymer layer.

As discussed above, the interaction between PNIPAM segments and the gold surface is strong, thus, PNIPAM chains are not only grafted with their end groups but also adsorbed with their segments on the gold surface. At a low grafting

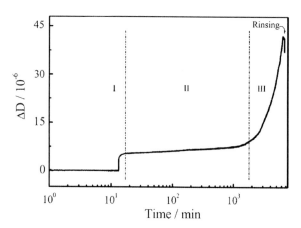

Fig. 3.3 Dissipation shift (ΔD) of the resonator immersed in the HS-PNIPAM solution as a function of logarithmic time, where the overtone number $n = 3$. Reprinted with the permission from Ref. [12]. Copyright 2005 American Chemical Society

density, the chains form a pancake-like structure. In regime II, as the grafting density increases, the uncovered area among the already grafted chains becomes narrow, the segments of the incoming chains are expected to be partially adsorbed on the surface, and other segments would turn up and overlap with the already existing chains. Because the uncovered area is still enough for accommodating an end group, the grafting would go along, but it becomes much slower. Meanwhile, the local segment–segment repulsion due to the local overlapping of the crowded chains has accumulated. Because the chemical bonding between the -HS groups and the gold surface is much stronger than the physical bonding of the segment-surface, as the result of the balance between the local segment–segment repulsion and the rubberlike elasticity of the chains, the absorbed segments begin to desorb and protrude from the gold surface, forming more loops and tails. The conformation of the grafted chains transits from a "pancake" to a "mushroom". After the rearrangement, the space among neighboring chains is large enough to accommodate the incoming chains, which makes further grafting possible. As the grafting density increases, the grafted chains form brushes from regime II to regime III. The grafting density will increase until the saturation is reached.

The conformation of the HS-PNIPAM chains can be clearly viewed in terms of the ΔD versus $-\Delta f$ relation shown in Fig. 3.4. The grafting involves two kinetic processes. The same ΔD versus $-\Delta f$ relation in regimes I and II indicates that there is not much difference between the conformations of the grafted chains in these two regimes. It is reasonable to ascribe the conformations in regimes I and II to a pancake and a mushroom, respectively, because the chains in both conformations are random coils. The slow increase in ΔD with $-\Delta f$ further indicates that the grafted chains have pancake and mushroom conformations. This is because the thickness of the layer formed by the grafted chains in such conformations only slightly increases with the grafting density. A much larger slope can be observed in regime III, indicating that the thickness increases with the grafting density more obviously than that in regimes I and II. Therefore, the grafted chains in regime III

3.2 Pancake-to-Brush Transition

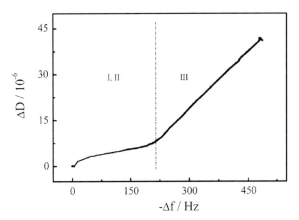

Fig. 3.4 ΔD versus $-\Delta f$ relation during the grafting of HS-PNIPAM chains, where the overtone number $n = 3$. Reprinted with the permission from Ref. [12]. Copyright 2005 American Chemical Society

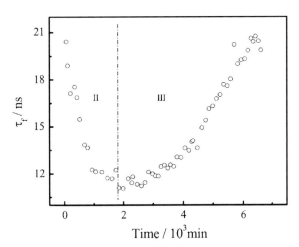

Fig. 3.5 Time dependence of characteristic relaxation time (τ_f) for the grafted PNIPAM chains fitted by using the Voigt model. Reprinted with the permission from Ref. [12]. Copyright 2005 American Chemical Society

have a more stretched conformation, i.e., the chains form brushes. Accordingly, from regime II to III, the mushroom-to-brush transition occurs.

Figure 3.5 shows the time dependence of characteristic relaxation time (τ_f) for the grafted HS-PNIPAM chains fitted by using the Voigt model. τ_f gradually decreases with time in regime II, but it increases in regime III until the grafting reaches a saturated state. It is known that the grafted chains form a pancake structure in the initial stage of regime II. PNIPAM segments are strongly bound to the gold surface, so that the motion of polymer chains is limited, leading to a large relaxation time. As the grafting density increases, some segments gradually desorb from the surface. The grafted chains become more mobile, so that the relaxation time gradually decreases. In regime III, though the segmental adsorption is very limited, further increasing the grafting density causes the degree of interpenetration between PNIPAM chains to increase, so that the relaxation time becomes longer again. Obviously, the mushroom-to-brush transition occurs at the minimum

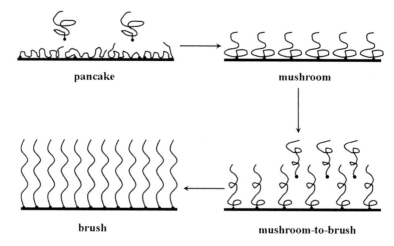

Fig. 3.6 Schematic illustration of the pancake-to-brush transition during the grafting of PNIPAM chains

relaxation time between regimes II and III. The schematic illustration of the pancake-to-brush transition is shown in Fig. 3.6.

3.3 Mushroom-to-Brush Transition

Actually, the three-regime-kinetics character can also be observed in the mushroom-to-brush transition [14]. A weak polyelectrolyte, poly[(2-dimethyl-amino)ethyl methacrylate] (PDEM), is employed to study the mushroom-to-brush transition. PDEM has a pK_a at ~7 [15]. Thus, PDEM chains will be completely charged, partially charged, and uncharged at pH 2, 6, and 10, respectively. Since the chain segment-surface interaction has significant effects on the kinetics and mechanism of the grafting of polymer chains, the interaction between PDEM segments and gold surface is examined by measuring the adsorption of dithioester-terminated PDEM (DTE-PDEM) on the gold-coated resonator surface. The only difference between DTE-PDEM and HS-PDEM is in the terminated group. Figure 3.7 shows the frequency shift of the resonator as a function of time after DTE-PDMEM was introduced at three different pH values. It can be seen that $\Delta f \sim 0$ at pH 2 and 6 after rinsing, indicating no segmental adsorption of DTE-PDEM on the gold surface. The fact also indicates that dithioester groups do not couple with gold surface. The frequency shift before rinsing arises from the changes of viscosity and density of the contacting medium. At pH 10, a small amount of DTE-PDEM chains are adsorbed on the gold surface, as reflected by a decrease of $\Delta f \sim 43$ Hz after rinsing, which indicates that some interaction exists between the uncharged segments and the gold surface, but the interaction is not strong. Thus, the grafted

3.3 Mushroom-to-Brush Transition

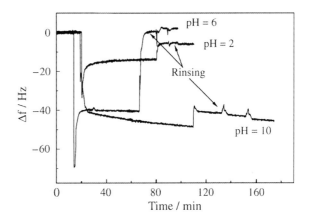

Fig. 3.7 Time dependence of frequency shift (Δf) of the resonator immersed in the DTE-PDEM solution, where the overtone number $n = 3$. Reprinted from Ref. [14], Copyright 2006, with permission from Elsevier

PDEM chains on the gold surface are expected to form a mushroom structure instead of a pancake structure at pH 2, 6, and 10 in the low grafting density regime.

Figure 3.8 shows the changes in Δf and ΔD of the resonator immersed in an aqueous solution at pH 10 as a function of logarithmic time after HS-PDEM was introduced. Obviously, the changes in Δf and ΔD show that the grafting of PDEM

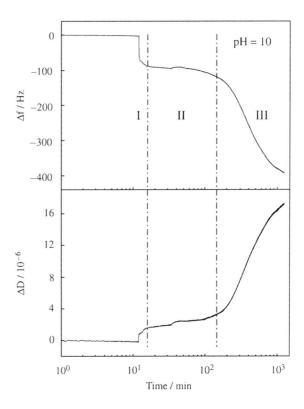

Fig. 3.8 The shifts in frequency (Δf) and dissipation (ΔD) of the resonator immersed in the HS-PDEM solution at pH 10 as a function of logarithmic time, where the overtone number $n = 3$. Reprinted from Ref. [14], Copyright 2006, with permission from Elsevier

chains has a three-regime-kinetics character. At the initial stage, the significant decrease in Δf (regime I) indicates that the chains quickly graft onto the bare gold surface. Subsequently, the grafting slows down (regime II) because the grafting is hindered by the already grafted chains on the surface. Finally, the grafting speeds up again, as reflected by the relatively sharp decrease in Δf (regime III), implying that the conformation of the already grafted chains has changed, so that the incoming chains can be grafted onto the surface. In parallel, the change in ΔD also gives the information on the structural change of the layer formed by the grafted chains. The quick increase in ΔD in regime I further indicates the grafting of the chains, whereas the slight increase in ΔD in regime II suggests that almost no grafting occurs. The large change in ΔD in regime III indicates the occurrence of the grafting again.

The changes in Δf and ΔD of the resonator immersed in the HS-PDEM solution at pH 6 and 2 are shown in Fig. 3.9. Like the case of the uncharged chains at pH 10, the grafting of either partially or completely charged chains also has a three-regime-kinetics character. Nonetheless, the span for regime II is strongly dependent on the degree of charging of the grafted chains. It covers ~ 130 min at pH 10 (Fig. 3.8), but it drops to ~ 90 min when the chains are partially charged at pH 6. The complete charging at pH 2 leads regime II to be only ~ 30 min. As discussed above, the grafting is very slow in regime II. What happens in regime II is the rearrangement of the grafted chains. Figures 3.8 and 3.9 clearly show that the time

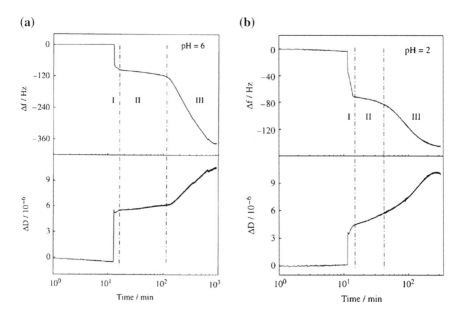

Fig. 3.9 The shifts in frequency (Δf) and dissipation (ΔD) of the resonator immersed in the HS-PDEM solution at pH 6 and 2 as a function of logarithmic time, where the overtone number $n = 3$. **a** pH 6, and **b** pH 2. Reprinted from Ref. [14], Copyright 2006, with permission from Elsevier

3.3 Mushroom-to-Brush Transition

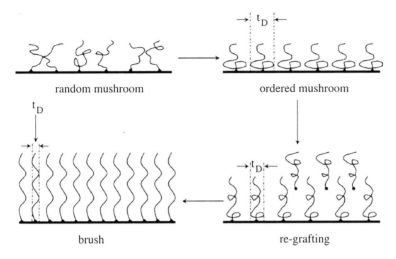

Fig. 3.10 Schematic illustration of the mushroom-to-brush transition during the grafting of PDEM chains. Reprinted from Ref. [14], Copyright 2006, with permission from Elsevier

required for the rearrangement decreases with the increasing degree of charging. In other words, more strongly charged chains can more easily rearrange themselves. However, why does such a rearrangement occur in regime II? The chains are quickly grafted in regime I, which is controlled by the centre-of-mass diffusion of the chains, causing the chains to be randomly tethered on the gold surface, and some local overlapping of the chains is resulted. Obviously, the locally overlapped chains are in a nonequilibrium state. Driven by the balance between the local segment–segment repulsion and the elasticity of the chains, the grafted chains tend to eliminate the local overlapping, and thus make a rearrangement by themselves. As the degree of charging increases, the electrostatic repulsion between the chains becomes stronger, and the chains tend to be more stretched with a smaller tube diameter [16] (Fig. 3.10), which leads the local overlapping to be more difficult. That is why the time required for the rearrangement decreases with the degree of charging. In short, the conformation of the grafted chains in regime II might be slightly different from that in regime I, i.e., the chains form a random mushroom in regime I but an ordered mushroom in regime II. The latter without local overlapping of the grafted chains makes further grafting possible. In other words, the space between two neighbor chains is enough to accommodate the incoming chains after the rearrangement.

The marked changes in Δf and ΔD from regime II to III suggest that the grafted chains may form brushes structure in regime III, which is much different from the mushroom structure in regimes I and II. The decrease in Δf and increase in ΔD with time in regime III are indicative of the increase of grafting density and the stretching of the grafted chains. Additionally, the absolute values for both Δf and ΔD at saturation decrease with the degree of charging. This is understandable. Due to the electrostatic repulsion, the incoming chain must keep its distance from the

Fig. 3.11 Plot of ΔD versus $-\Delta f$ relation for the grafting of HS-PDEM chains at pH 2, 6, and 10, where the overtone number $n = 3$. Reprinted from Ref. [14], Copyright 2006, with permission from Elsevier

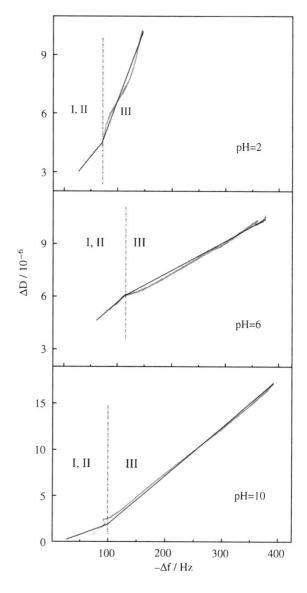

already grafted chains so that it can be grafted. A stronger repulsion would lead to a larger distance and a smaller grafting density. Thus, Δf and ΔD at saturation, respectively, decreases and increases with the increase of pH.

The grafting kinetics can be better viewed in terms of ΔD versus $-\Delta f$ relation in Fig. 3.11. Clearly, the grafting for uncharged, partially charged, and completely charged chains involves two kinetic processes. Regimes I and II have the same ΔD versus $-\Delta f$ relation, indicating only a minor difference between the conformations of the grafted chains in these two regimes; namely, the grafted chains form a mushroom structure in both regimes I and II. In regime III, a quite different kinetic process can be observed, suggesting that the grafted chains in this regime have a conformation different from that in regimes I and II, i.e., the grafted chains form brushes. Accordingly, from regime II to III, the mushroom-to-brush transition occurs.

3.4 Conclusion

The QCM-D results reveal that the competition between the segment–segment repulsion and the segment-surface attraction plays a crucial role in the formation of polymer brushes. For the HS-PNIPAM chains whose chain segments strongly interact with the gold surface, the grafted chains undergo a pancake-to-mushroom-to-brush transition with a three-regime-kinetics character during the grafting. For the HS-PDEM chains, a three-regime-kinetics character is also observed during the grafting. Due to the weak segment-surface interaction, HS-PDEM chains form a random mushroom structure in regime I. In regime II, the chains rearrange themselves to eliminate the local overlapping and form an ordered mushroom conformation. In regime III, the chains form brushes. For either HS-PNIPAM or HS-PDEM, the mushroom-to-brush transition occurs in the region from regime II to III.

References

1. Halperin A, Tirrell M, Lodge TP (1992) Tethered chains in polymer microstructures. Adv Polym Sci 100:31–71
2. Azzaroni O (2012) Polymer brushes here, there, and everywhere: Recent advances in their practical applications and emerging opportunities in multiple research fields. J Polym Sci Polym Chem 50:3225–3258
3. Ligoure C, Leibler L (1990) Thermodynamics and kinetics of grafting end-functionalized polymers to an interface. J Phys-Paris 51:1313–1328
4. Hasegawa R, Doi M (1997) Dynamical mean field calculation of grafting reaction of end-functionalized polymer. Macromolecules 30:5490-5493
5. Abraham T, Giasson S, Gohy JF, Jerome R, Muller B, Stamm M (2000) Adsorption kinetics of a hydrophobic-hydrophilic diblock polyelectrolyte at the solid-aqueous solution interface: a slow birth and fast growth process. Macromolecules 33:6051–6059
6. Penn LS, Huang H, Sindkhedkar MD, Rankin SE, Chittenden K, Quirk RP, Mathers RT, Lee Y (2002) Formation of tethered nanolayers: three regimes of kinetics. Macromolecules 35:7054–7066
7. Huang H, Penn LS, Quirk RP, Cheong TH (2004) Effect of segmental adsorption on the tethering of end-functionalized polymer chains. Macromolecules 37:516–523
8. Huang HQ, Rankin SE, Penn LS, Quirk RP, Cheong TH (2004) Transition from mushroom to brush during formation of a tethered layer. Langmuir 20:5770–5775
9. Alexander S (1977) Adsorption of chain molecules with a polar head a-scaling description. J Phys-Paris 38:983–987

10. Ligoure C (1993) Polymers at interfaces—from a quasi self-similar adsorbed layer to a quasi brush first-order phase-transition. J Phys II 3:1607–1617
11. Zhu MQ, Wang LQ, Exarhos GJ, Li ADQ (2004) Thermosensitive gold nanoparticles. J Am Chem Soc 126:2656–2657
12. Liu GM, Cheng H, Yan LF, Zhang GZ (2005) Study of the kinetics of the pancake-to-brush transition of poly(N-isopropylacrylamide) chains. J Phys Chem B 109:22603–22607
13. Love JC, Estroff LA, Kriebel JK, Nuzzo RG, Whitesides GM (2005) Self-assembled monolayers of thiolates on metals as a form of nanotechnology. Chem Rev 105:1103–1169
14. Liu GM, Yan LF, Chen X, Zhang GZ (2006) Study of the kinetics of mushroom-to-brush transition of charged polymer chains. Polymer 47:3157–3163
15. Lee AS, Gast AP, Butun V, Armes SP (1999) Characterizing the structure of pH dependent polyelectrolyte block copolymer micelles. Macromolecules 32:4302–4310
16. de Gennes PG (1979) Scaling Concepts in Polymer Dynamics. Ithaca, New York

Chapter 4
Growth Mechanism of Polyelectrolyte Multilayers

Abstract In this chapter, QCM-D is employed to systematically study the influences of temperature, pH, salt concentration, salt type, chain flexibility, and chain architecture on the growth of polyelectrolyte multilayers. In the case of sodium poly(styrene sulfonate)/poly[2-(dimethylamino)ethyl methacrylate] multilayer, the multilayer growth is dominated by the chain interpenetration which can be modulated by varying temperature, pH, and salt concentration. In the case of sodium poly(styrene sulfonate)/poly(diallyldimethylammonium chloride) multilayer, the multilayer growth is dominated by chain conformation and chain interpenetration at $C_{NaCl} < 1.0$ M and $C_{NaCl} \geq 1.0$ M, respectively. The specific ion effect on the growth of polyelectrolyte multilayers can be observed in water, methanol, as well as their mixtures, and the ion specificity is determined by the specific interactions between the charged groups and the counterions. When the multilayer is constructed by two semiflexible polyelectrolytes, the multilayer growth is controlled by the delicate balance between the weakening of electrostatic repulsion between the identically charged groups on the same chain and the decrease of electrostatic attraction between the neighboring layers with the increase of salt concentration. The influence of arm number on the chain interpenetration during the multilayer growth is dominated by the steric effect created by the arm chains.

Keywords Chain interpenetration · Chain conformation · Layer-by-layer · Electrostatic interaction · Specific ion effect · Ion pair · Chain rigidity · Chain architecture

4.1 Introduction

The sequential layer-by-layer (LbL) deposition of oppositely charged polyelectrolytes on a solid substrate can generate a polyelectrolyte multilayer (PEM) [1–3]. Because of the potential applications of PEMs in various fields, the growth of PEMs has been investigated extensively [4–7]. In general, the formation of PEMs is influenced by temperature [8–10], pH [11–13], ionic strength [14–16], ion type [17–20], solvent quality [21–23], molecular weight [24, 25], chain rigidity [26–28], and chain architecture [29–31]. However, the exact mechanism of how the growth of PEMs is influenced by the external conditions and the intrinsic properties of polyelectrolyte chains still remains elusive. For example, it is reported that the thickness of PEM increases with temperature. One explanation is that more chains are deposited due to the fact that the hydrophobic interaction increases as the temperature increases [9]. Alternatively, Tan et al. [10] suggest that PEM is more swollen at an elevated temperature, which causes more polyelectrolyte chains to be trapped in the deposited layers, leading to an increase of the thickness. With regard to the pH effect, some studies show that the deposition of weak polyelectrolytes is dominated by chain charge density [11, 12], while other studies reveal that the matching of charge densities between two polyelectrolytes plays a critical role in the multilayer growth [13]. For the specific ion effect on the growth of PEMs, Salomaki et al. [17] show that the specific anion effect on the thickness, the storage shear modulus, and the swelling extent of PEM is related to the hydration entropy of anions. Dubas and Schlenoff [20] suggest that the ion-specific growth of PEM is correlated with hydrophobicity and affinity of counterions.

On the other hand, the multilayer growth usually exhibits two different modes, i.e., linear and exponential growth modes [14, 32]. In the former case, the mass and thickness linearly increase with the layer number [14]. In contrast, the mass and thickness exponentially increase with the layer number in the latter case [32]. By varying salt concentration [33], temperature [34], solvent quality [21], and molar ratio of polyelectrolytes [35], the mode of multilayer growth can be changed from linear to exponential. Understanding the mechanism of such a transition is important for controlling the construction of PEMs. The LbL buildup has been predicted to be inherently exponential by a theoretical model, but the film will grow linearly when the polyelectrolyte chains do not diffuse fast enough within the multilayer during the deposition [34]. It is thought that the chain interpenetration of polyelectrolytes is vital in multilayer growth for both linear and nonlinear growth modes [32, 36]. More specifically, the former occurs when no polyelectrolyte chains diffuse within the multilayer, but the latter happens with the chain diffusion throughout the multilayer [32, 36].

The chain conformation is also thought to have significant influences on the multilayer growth because the extent of surface charge overcompensation is governed by the chain conformation on the surface [37]. Moreover, the chain interpenetration is also related to the chain conformation; for instance, a more coiled chain conformation can lead to a higher extent of chain interpenetration [8, 16]. Generally, the chain conformation of polyelectrolyte is determined not only by the chain intrinsic properties, but also by the external conditions. In this chapter, QCM-D is used to systematically study the growth of PEMs as functions of external conditions and intrinsic properties of polyelectrolyte chains. We will find that QCM-D can provide not only the information on the changes in mass, thickness, and structure of the PEMs but also the information on the changes in

4.1 Introduction

conformation and interpenetration of polyelectrolyte chains within the PEMs, which will be useful to clarify the growth mechanism of PEMs.

4.2 Roles of Chain Interpenetration and Conformation in the Growth of PEMs

The typical shifts in Δf and ΔD for the LbL deposition of sodium poly(styrene sulfonate)/poly[2-(dimethylamino)ethyl methacrylate] (PSS/PDEM) are shown in Fig. 4.1, where pH is fixed at 4, ionic strength is fixed at 0.2 M, and T is of 20 °C. A measurement of LbL deposition is initiated by switching the liquid exposed to the resonator from water to a poly(ethylene imine) (PEI) solution with a polymer concentration of 1.0 mg/mL. PEI is allowed to adsorb onto the resonator surface for ∼20 min before the surface is rinsed with water to ensure a uniform coating with positive charges, so that the effects of substrate on the growth of multilayer are minimized [21]. After water is replaced with phosphate buffer, 0.1 mg/mL PSS and PDEM are alternately introduced for ∼20 min with buffer rinsing in between in case the polyelectrolytes form complexes in the solution. The decrease in Δf and the increase in ΔD clearly indicate that the polyelectrolyte chains gradually deposit onto the resonator surface driven by the electrostatic interaction.

Figure 4.2 shows the shifts in $-\Delta f$ and ΔD as a function of layer number for the growth of PSS/PDEM multilayer at pH 4, where T is of 20, 25, and 30 °C. PSS is a strong polyelectrolyte whose charge degree is independent of pH. In contrast, PDEM is a weak polyelectrolyte with $pK_a \sim 7$ [38]. Thus, PDEM is completely charged at pH 4. Accordingly, the LbL deposition of PSS/PDEM at pH 4 deals with two completely charged polyelectrolytes. In Fig. 4.2, both $-\Delta f$ and ΔD linearly increase with layer number, that is, the PEM exhibits a linear growth manner. The increase of $-\Delta f$ indicates that the two oppositely charged polyelectrolytes alternatively deposit on the resonator surface. The increase of ΔD further indicates the sequential deposition of polyelectrolytes. For the same layer number, $-\Delta f$ increases with

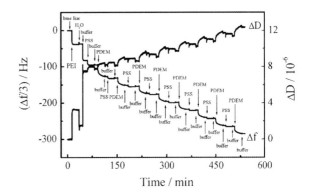

Fig. 4.1 Shifts in frequency (Δf) and dissipation (ΔD) for the LbL deposition of PSS/PDEM at pH 4 and T of 20 °C, where the overtone number $n = 3$. Reprinted with permission from Ref. [8]. Copyright 2008 American Chemical Society

Fig. 4.2 Shifts in frequency ($-\Delta f$) and dissipation (ΔD) as a function of layer number for the growth of PSS/PDEM multilayer at pH 4 and T of 20, 25, and 30 °C, where the overtone number $n = 3$. The odd and even layer numbers correspond to the deposition of PSS and PDEM, respectively. Reprinted with permission from Ref. [8]. Copyright 2008 American Chemical Society

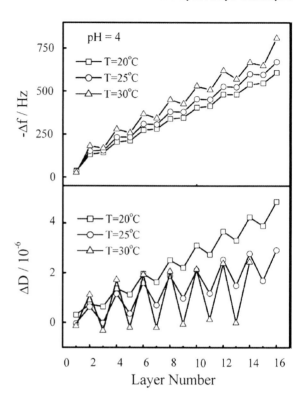

temperature, suggesting that the elevated temperature facilitates the deposition of polyelectrolytes.

The most important event is that $-\Delta f$ exhibits an oscillation character, namely, $-\Delta f$ oscillatedly increases with the layer number. So does ΔD. However, ΔD exhibits a more remarkable oscillation. Because the dissipation factor of a film relates to its structure, the increase of ΔD from odd to even layer number implies that the layer changes from dense to loose due to the deposition of a swollen PDEM layer. Then, the complexation between PSS and PDEM occurs after PSS is added. Such a complexation could extend a certain depth in the layer, that is, the chain interpenetration between PSS and PDEM layers occurs. As a result, the layer becomes denser, and ΔD drops. At the same time, the surface charge is changed from positive to negative, which makes the subsequent adsorption of PDEM chains possible. Consequently, the alternative swelling-and-shrinking of the outermost layer leads to the oscillations of ΔD. On the other hand, $-\Delta f$ first increases after PDEM is introduced due to the increase in thickness of the layer. Then, the layer shrinks after PSS is introduced due to the complexation. This gives rise to a decrease in thickness, so $-\Delta f$ decreases. Clearly, the oscillation in $-\Delta f$ also results from the alternative swelling-and-shrinking of the outermost layer.

Because the interpenetration increases with the degree of complexation between polyelectrolytes, the extent of interpenetration can be viewed in terms of

Fig. 4.3 Shifts in frequency ($-\Delta f$) and dissipation (ΔD) as a function of layer number for the growth of PSS/PDEM multilayer at T of 25 °C and pH of 4, 7, and 10, where the overtone number $n = 3$. The odd and even layer numbers correspond to the deposition of PSS and PDEM, respectively. Reprinted with permission from Ref. [8]. Copyright 2008 American Chemical Society

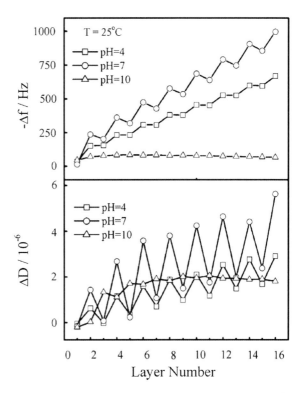

the amplitude of oscillation in $-\Delta f$ and ΔD. In other words, larger amplitudes in $-\Delta f$ and ΔD imply more interpenetration. It has been suggested that a linearly growing multilayer exhibits a somewhat fuzzy but layered structure due to the chain interpenetration between neighboring layers [39]. In Fig. 4.2, the oscillation amplitude of either $-\Delta f$ or ΔD increases with temperature, indicating that the interpenetration is strongly dependent on temperature. This is because the polyelectrolyte chains are more fluid and the multilayer is more swollen at an elevated temperature [10], which enhances the polyelectrolyte complexation and interpenetration. Thus, the elevated temperature facilitates the deposition of polyelectrolytes. In addition, ΔD has almost the same amplitude at the same temperature for different layer numbers, indicating that the interpenetration between any neighboring layers is equal. This might be the reason that $-\Delta f$ increases linearly with the layer number.

The electrostatic interaction and chain conformation of a weak polyelectrolyte are affected by its charge density which can be tuned by pH. Thus, the growth of PSS/PDEM multilayer is also expected to be influenced by pH. Figure 4.3 shows the changes in $-\Delta f$ and ΔD as a function of layer number for the growth of PSS/PDEM multilayer at 25 °C, where pH is of 4, 7, and 10. At pH 10, $-\Delta f$ keeps almost constant with layer number, indicating that no multilayer is formed. This is because PDEM is uncharged at pH 10 and no electrostatic interaction exists

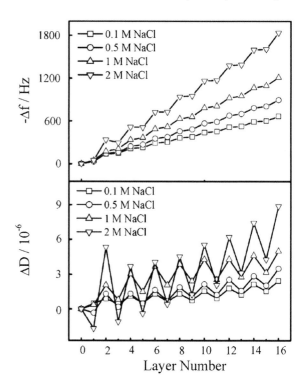

Fig. 4.4 Shifts in frequency ($-\Delta f$) and dissipation (ΔD) as a function of layer number for the growth of PSS/PDEM multilayer at T of 20 °C and pH 4 with different NaCl concentrations, where the overtone number n = 3. The odd and even layer numbers correspond to the deposition of PSS and PDEM, respectively. Reprinted with permission from Ref. [8]. Copyright 2008 American Chemical Society

between PSS and PDEM at this pH. For the same layer number, $-\Delta f$ increases with pH from 4 to 7. It is known that the charge density of PDEM chain decreases with the increasing pH [38]. As the charge density decreases, more polyelectrolyte chains are adsorbed to overcompensate and invert the surface charge so that the adsorption of next layer occurs. As a result, $-\Delta f$ increases with the decreasing charge density for the same layer number. Also, the amplitude of oscillation in $-\Delta f$ or ΔD increases as pH increases from 4 to 7. Similar to the case of temperature effect, this fact is indicative of the increase of extent of interpenetration. The decrease of charge density with pH causes the polyelectrolyte chains to adopt a more coiled conformation, and the multilayer becomes more swollen, which makes more interpenetration possible. At pH 10, $-\Delta f$ and ΔD do not exhibit obvious oscillation, indicating that no interpenetration occurs in the deposition due to the lack of electrostatic interaction.

As discussed above, the effects of temperature and pH can be attributed to the variation of interpenetration between the neighboring layers. To further extract its role, the salt effect on the LbL deposition of PSS/PDEM has also been examined (Fig. 4.4). For the same layer number, the increases of $-\Delta f$ and ΔD with the NaCl concentration (C_{NaCl}), indicating that the thickness of PEM increases with the salt concentration. Meanwhile, the extent of interpenetration also increases with C_{NaCl}, as indicated by the fact that the amplitude of oscillation in ΔD increases with

4.2 Roles of Chain Interpenetration and Conformation in the Growth of PEMs

C_{NaCl}. Dubas and Schlenoff [20] suggested that the thickness increment with the salt concentration is mainly determined by the charge penetration length. Since the multilayer at a higher salt concentration becomes more swollen, the oppositely charged polyelectrolyte chains would diffuse into the interior of the multilayer. Consequently, more polyelectrolytes are trapped in the multilayer, the charge penetration length increases, and the layer thickness increases. This is why $-\Delta f$ increases with salt concentration for the same layer number in Fig. 4.4. On the other hand, the amplitude of oscillation in ΔD increases with salt concentration, further indicating an increase of the extent of interpenetration. In short, the growth of PSS/PDEM multilayer is dominated by the interpenetration between neighboring layers.

In the case of PSS/PDEM multilayer, the growth of PEM always exhibits a linear manner regardless of temperature, pH, and salt concentration, which might be determined by the intrinsic properties of the polyelectrolytes. However, the growth of PEM may exhibit an exponential manner at high salt concentrations for other systems, e.g., the sodium poly(styrene sulfonate)/poly(diallyldimethylammonium chloride) (PSS/PDDA) multilayer [16]. In Fig. 4.5, $-\Delta f$ increases with C_{NaCl} for the same layer number. Meanwhile, $-\Delta f$ exhibits a more obvious exponential character at a higher C_{NaCl}. In contrast, ΔD only slightly increases with layer number and does not have dependence on salt concentration at $C_{NaCl} < 1.0$ M. ΔD gradually increases with layer number at C_{NaCl} of 2.0 M and remarkably increases at C_{NaCl} of 3.0 M. The facts indicate that the mechanism for the growth of PSS/PDDA multilayer in the range of $C_{NaCl} < 1.0$ M might be different from that at $C_{NaCl} > 1.0$ M. Also, ΔD shows oscillation at C_{NaCl} of 2.0 and 3.0 M and the amplitude gradually increases with layer number. The oscillation reflects the interpenetration and complexation of the polyelectrolytes. For a certain PSS outer layer, when PDDA is introduced, a swollen PDDA layer is formed on the PSS surface. Thus, ΔD increases. However, the subsequently introduced PSS chains would penetrate into the PDDA layer, leading to the complexation between PSS and PDDA chains. As a result, the layer becomes denser, as reflected by the drop in ΔD. At the same time, the surface charge changes from positive to negative, making the subsequent adsorption of PDDA chains on the surface possible. The layer number dependence of amplitude at C_{NaCl} of 2.0 and 3.0 M indicates the enhancement of interpenetration with the layer number.

A better view of the layer number dependence of ΔD in the range of $C_{NaCl} \leq 1.0$ M is shown in the inset of Fig. 4.5. When $C_{NaCl} \leq 0.05$ M, no interpenetration occurs since ΔD does not have an oscillation. Nevertheless, for the same layer number, ΔD at C_{NaCl} of 0.05 M is larger than that in a salt-free solution. This is because a more swollen film is resulted in the presence of NaCl. In the range of 0.05 M $< C_{NaCl} < 1.0$ M, an obvious oscillation in ΔD can be observed, which is indicative of the occurrence of the interpenetration. The amplitude of oscillation is independent of layer number, suggesting that the interpenetration in each layer has the same degree. Moreover, the amplitude is also independent on C_{NaCl}, implying that the addition of salt does not influence the interpenetration. Obviously, the multilayer growth is not dominated by the

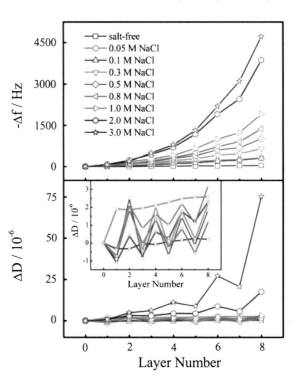

Fig. 4.5 Layer number dependence of shifts in frequency ($-\Delta f$) and dissipation (ΔD) as a function of NaCl concentration (C_{NaCl}) for the growth of PSS/PDDA multilayer, where the overtone number $n = 3$. The inset shows the change in ΔD for the multilayer growth as a function of layer number in the range of $C_{NaCl} \leq 1.0$ M. The odd and even layer numbers correspond to the deposition of PSS and PDDA, respectively. Reprinted with permission from Ref. [16]. Copyright 2008 American Chemical Society

interpenetration in this range of C_{NaCl}. Otherwise, $-\Delta f$ should exhibit a linear growth regardless of the salt concentration. In addition, the amplitude at C_{NaCl} of 1.0 M shows a gradual increase with layer number, indicating that 1.0 M is close to a critical salt concentration for the multilayer growth from one regime to another.

Atomic force microscopy investigations have shown that the buildup regime from linear to exponential is due to the continuous increase of active surface area available for adsorption. This is because the polyelectrolyte chains change from an extended to a more compact conformation as the salt concentration increases [40]. In parallel, the exponential growth is suggested to be attributed to the "in" and "out" diffusion of polyelectrolyte chains through the multilayer during the buildup [32, 36]. To further examine the effect of NaCl concentration on the growth of PSS/PDDA multilayer, the values of Δf with layer number at different NaCl concentrations in Fig. 4.5 are fit based on the following equation [33]:

$$\Delta f = A \exp(\alpha N) + B \tag{4.1}$$

where A and B are two constants, and α and N are the characteristic growth parameter and layer number, respectively.

Figure 4.6 shows α gradually increases from 0 to ~ 0.4 as C_{NaCl} increases from 0 to 3.0 M. Thus, the exponential growth mode gradually becomes dominant as

4.2 Roles of Chain Interpenetration and Conformation in the Growth of PEMs

C_{NaCl} increases. Generally, the conformation of polyelectrolytes is determined by the electrostatic interaction, which can be described in terms of Debye length (l_D). The decrease of l_D with C_{NaCl} would screen the electrostatic interaction so that the polyelectrolyte chains would adopt a more coiled conformation. Thus, if the LbL deposition is dominated by the chain conformation, the characteristic growth parameter should relate to l_D. Interestingly, the fitting curve and the experimental data are superposed well in the range of $C_{NaCl} < 1.0$ M, implying that the multilayer growth is dominated by the chain conformation. In the range of $C_{NaCl} \geq 1.0$ M, the fitting curve gradually deviates from the experimental data, indicating that the LbL deposition is no longer dominated by the chain conformation.

The combination of the results in Figs. 4.5 and 4.6 indicates that the multilayer growth in the range of $C_{NaCl} < 1.0$ M is dominated by the chain conformation instead of the chain interpenetration. In the range of $C_{NaCl} \leq 0.05$ M, a higher salt concentration leads to a more swollen film without interpenetration. In the range of 0.05 M $< C_{NaCl} < 1.0$ M, more loops and tails are formed at the polymer/solution interface as C_{NaCl} increases, so that the multilayer has a higher charge overcompensation level. This is why $-\Delta f$ increases with salt concentration for the same layer number. Moreover, such coils and loops would lead the active surface area available for adsorption to increase continuously. Consequently, $-\Delta f$ exhibits an exponential increase with layer number, particularly at the high salt concentrations.

At $C_{NaCl} \geq 1.0$ M, the fitting curve gradually deviates from the experimental data, indicating that the multilayer growth is no longer governed by the chain conformation. This is because the charge overcompensation attains a critical level at a certain salt concentration due to the electrostatic repulsion between the like charges. Further addition of the salt causes the oppositely charged polyelectrolyte chains to diffuse into the interior of the multilayer as the mobility of the adsorbed polyelectrolyte chains increases with the salt concentration [34]. Therefore, a higher salt concentration gives rise to a thicker film and a larger value of $-\Delta f$. On

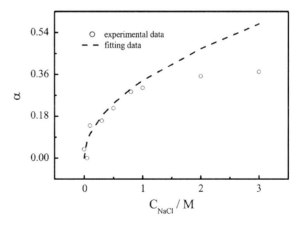

Fig. 4.6 NaCl concentration (C_{NaCl}) dependence of characteristic growth parameter (α), where the fitting is based on $\alpha = \beta/l_D$; l_D is the Debye length and β is fixed at 0.1. Reprinted with permission from Ref. [16]. Copyright 2008 American Chemical Society

the other hand, the amplitude of oscillation in ΔD increases with layer number, indicating an increase of the extent of interpenetration, which would lead the multilayer to grow acceleratedly in the early stage of growth because the interpenetration is limited by the impenetrable interface [41]. Actually, this mechanism also involves the mode of "in" and "out" diffusion of polyelectrolytes during the LbL deposition [32, 36].

4.3 Specific Ion Effect on the Growth of PEMs

In fact, the growth of PEMs is influenced not only by temperature, pH, and ionic strength, but also by ion type. In general, ions can be categorized as kosmotropes and chaotropes in light of the strength of ionic hydration [42]. The strongly hydrated ions are usually defined as kosmotropes, whereas the weakly hydrated ions are called chaotropes [43]. Collins has proposed a concept that only oppositely charged ions with similar water affinities can form strong ion pairs, which dominates the ion-specific interactions in aqueous solutions [44]. Parsons and Ninham [45] have suggested that specific ion effect is due to the polarizability of ions and is manifested through the ionic dispersion forces. Figure 4.7 shows the layer number dependence of $-\Delta f$ and ΔD for the growth of PSS/PDDA multilayer as a function of salt type. $-\Delta f$ increases with layer number, indicating the sequential deposition of polyelectrolytes. For the same layer number, $-\Delta f$ increases following the order $SO_4^{2-} < H_2PO_4^- < CH_3COO^- < F^- < HCO_3^- < Cl^- < ClO_3^- < Br^-$, which is roughly consistent with the classical Hofmeister series [42]. Hence, the deposition of polyelectrolytes is affected by the nature of anions. Furthermore, the anions can be divided into two groups with HCO_3^- as the borderline. For the cases of SO_4^{2-}, $H_2PO_4^-$, CH_3COO^-, F^- and HCO_3^-, $-\Delta f$ linearly increases with layer number. In contrast, Cl^-, ClO_3^-, and Br^- cause the multilayer to grow nonlinearly. On the other hand, ΔD gradually increases with the layer number, further indicating the sequential deposition of polyelectrolytes. However, ΔD only has slight dependence on anion species except in the cases of Br^-, HCO_3^-, and SO_4^{2-}. The relatively low ΔD observed in Na_2SO_4 solution indicates that SO_4^{2-} leads to a thin and rigid multilayer, whereas the larger ΔD observed in NaBr solution reflects a thicker and more swollen multilayer resulted. The multilayer deposited in $NaHCO_3$ solution also exhibits a remarkable increase in ΔD as layer number increases. To clarify the anion-specificity, the specific anion effect on the multilayer growth in linear and nonlinear modes is discussed separately.

Figure 4.8 shows the average value of $-\Delta f$ caused by the deposition of PDDA for the 14 and 16th layers gradually increases along the series $HCO_3^- < Cl^- < ClO_3^- < Br^-$, whereas the average value of $-\Delta f$ induced by the deposition of PSS for the 13 and 15th layers gradually decreases following the series $HCO_3^- > Cl^- > ClO_3^- > Br^-$. For the case of HCO_3^-, $-\Delta f$ values caused by the deposition of PSS and PDDA are almost equal to each other. For a certain

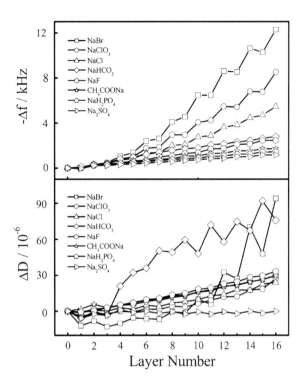

Fig. 4.7 Shifts in frequency ($-\Delta f$) and dissipation (ΔD) as a function of layer number for the growth of PSS/PDDA multilayer for different anions with common cation Na$^+$, where the overtone number $n = 3$ and the ionic strength is fixed at 0.5 M. The odd and even layer numbers correspond to the deposition of PSS and PDDA, respectively. Reprinted with permission from Ref. [19]. Copyright 2010 American Chemical Society

PSS outer layer, when PDDA is introduced, it forms a layer on PSS surface via the electrostatic attraction. Thus, the mass or layer thickness increases, leading to an increase of $-\Delta f$. However, the subsequently introduced PSS chains would penetrate into PDDA layer, and they form complexes. Some associated water molecules are released from the multilayer during the complexation, giving rise to a decrease in $-\Delta f$. Accordingly, more chain interpenetration and complexation

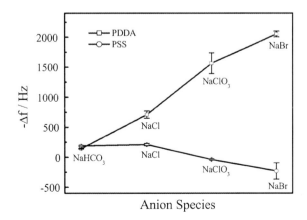

Fig. 4.8 The average frequency shift ($-\Delta f$) due to the deposition of PSS or PDDA for the last four layers as a function of anion species regarding Br$^-$, ClO$_3^-$, Cl$^-$ and HCO$_3^-$. Reprinted with permission from Ref. [19]. Copyright 2010 American Chemical Society

would cause more water molecules to release out, so that a larger decrease in $-\Delta f$ can be observed. The fact that $-\Delta f$ for the deposition of PSS decreases from HCO_3^- to Br^- indicates that the degree of chain interpenetration increases following the series. When PDDA is introduced to PSS surface again, the penetrated PSS chains may diffuse out and interact with PDDA chains forming a new layer [32]. Therefore, more PSS penetration results in more adsorption of PDDA, that is, a smaller $-\Delta f$ for the deposition of PSS gives rise to a larger $-\Delta f$ for the deposition of PDDA. In other words, the nonlinear growth of PSS/PDDA multilayer is dominated by the anion-modulated chain interpenetration. According to the law of matching water affinities [43], a more chaotropic anion can form a stronger ion pair with the weakly hydrated ammonium group, so that a more chaotropic anion can screen the polyelectrolyte charges more effectively, leading to a higher level of "extrinsic charge compensation" [41]. Therefore, as the anions change from Cl^- to Br^- along the series, a more swollen PDDA layer is resulted, which will facilitate the PSS chain penetration. Besides, the similar values of $-\Delta f$ for the deposition of PSS and PDDA in the $NaHCO_3$ solution suggests that only slight chain interpenetration occurs between the neighboring layers. Two oppositely charged polyelectrolytes with a high level of chain interpenetration and complexation would result in a rigid multilayer, whereas those with a low level of chain interpenetration are expected to form a loose multilayer. This is why the multilayer deposited in $NaHCO_3$ solution exhibits a dramatic increase in ΔD as layer number increases (Fig. 4.7).

The average values of $-\Delta f$ due to the deposition of PSS or PDDA for the last four layers as a function of anion species for HCO_3^-, F^-, CH_3COO^-, $H_2PO_4^-$, and SO_4^{2-} are shown in Fig. 4.9. It is evident that the deposition of PSS causes more increase in $-\Delta f$ than that of PDDA with the exception of the case of HCO_3^-, which implies that PSS chains form a swollen layer on PDDA surface. The subsequently introduced PDDA chains penetrate into such a swollen layer and form complexes with PSS chains. This is quite different from that for the nonlinear

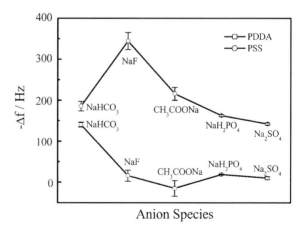

Fig. 4.9 The average frequency shift $(-\Delta f)$ due to the deposition of PSS or PDDA for the last four layers as a function of anion species regarding HCO_3^-, F^-, CH_3COO^-, $H_2PO_4^-$, and SO_4^{2-}. Reprinted with permission from Ref. [19]. Copyright 2010 American Chemical Society

4.3 Specific Ion Effect on the Growth of PEMs

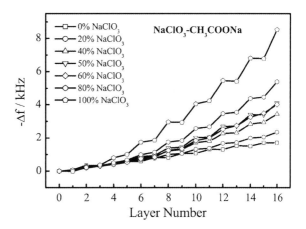

Fig. 4.10 Shift in frequency ($-\Delta f$) as a function of layer number for the growth of PSS/PDDA multilayer in the $NaClO_3$–CH_3COONa mixed solutions, where the overtone number $n = 3$ and the ionic strength is fixed at 0.5 M. The odd and even layer numbers correspond to the deposition of PSS and PDDA, respectively. Reprinted with permission from Ref. [19]. Copyright 2010 American Chemical Society

growth of PSS/PDDA multilayer described above. The $-\Delta f$ induced by the deposition of PDDA decreases from HCO_3^- to F^-, and then holds almost constant at ~ 0 from F^- to SO_4^{2-}. On the other hand, $-\Delta f$ induced by the deposition of PSS increases from HCO_3^- to F^-, followed by a gradual decrease of $-\Delta f$ from F^- to SO_4^{2-}. The increase of $-\Delta f$ for PSS and the decrease of $-\Delta f$ for PDDA from HCO_3^- to F^- suggest that the multilayer growth changes from a PSS penetration-dominated regime to a PDDA penetration-dominated regime. From F^- to SO_4^{2-}, $-\Delta f$ for the deposition of PDDA keeps almost constant, indicating that PDDA exhibits a similar level of penetration in the presence of different anions. Therefore, the linear multilayer growth is not dominated by the chain interpenetration. $-\Delta f$ for the deposition of PSS gradually decreases from F^- to SO_4^{2-}, indicating a gradual decrease of the amount of adsorbed PSS chains. Since ammonium is a weakly hydrated group, the effectiveness of charge screening by the anions should decrease from chaotropes to kosmotropes according to the concept of matching water affinities [43]. Therefore, the effectiveness of kosmotropic anions to screen the polyelectrolyte charges increases from SO_4^{2-} to F^-. At a more screened PDDA surface, PDDA chains would adopt a more loopy conformation, yielding a higher surface charge density. As a result, the PDDA surface can adsorb more subsequently introduced PSS chains with a larger $-\Delta f$. Clearly, the linear multilayer growth is dominated by the anion-modulated conformation of PDDA chains on the surface.

Although the specific ion effect is usually observed in mixed electrolyte systems such as biological systems, one always likes to construct PEMs in a single electrolyte solution instead of a mixed electrolyte solution for simplification. Since chaotropes and kosmotropes exhibit different interactions with the charged groups on polyelectrolyte chains, the ion specificity in a mixed electrolyte solution containing both chaotropes and kosmotropes is expected to be different from that in a single electrolyte solution. Figure 4.10 shows the layer number dependence of $-\Delta f$ in the $NaClO_3$–CH_3COONa mixed solutions. It can be seen that the multilayer

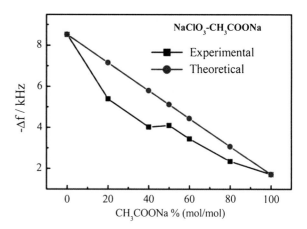

Fig. 4.11 Comparison between experimental and theoretical values of the frequency shift ($-\Delta f$) for the eight bilayers of PSS/PDDA multilayer in the $NaClO_3$-CH_3COONa mixed solutions as a function of molar fraction of CH_3COONa. Reprinted with permission from Ref. [19]. Copyright 2010 American Chemical Society

growth is gradually dominated by the nonlinear mode with the increasing molar fraction of $NaClO_3$. The deposition of PDDA chains causes more increase in $-\Delta f$ than PSS chains in the solution of 50 % $NaClO_3$ (mol/mol), similar to the deposition in chaotropic anion solutions. Therefore, the chaotropic anions have stronger influences on the multilayer growth than the kosmotropic anions in the mixed electrolyte solution.

If there is no interplay between the chaotropic anion-polyelectrolyte interaction and the kosmotropic anion-polyelectrolyte interaction in the mixed electrolyte solutions, the resulted $-\Delta f$ by the deposition of multilayer would follow the additivity law. However, Fig. 4.11 shows that the resulted $-\Delta f$ for the eight bilayers are always less than the theoretical values obtained on the basis of the additivity law. In other words, the specific anion effect on the multilayer growth in the mixed electrolyte solutions is nonadditive and an anion competition effect might occur there [46]. Since the weakly hydrated chaotropic anion can form strong ion pair with the weakly hydrated ammonium group on PDDA chains, the chaotropic anions would bind onto the polyelectrolyte chains more tightly in comparison with the kosmotropic anions. In the mixed electrolyte solutions, the chaotropic anions should prefer to bind onto the polyelectrolyte chains. Moreover, they may also replace the already adsorbed kosmotropic anions. Thus, the anion competition effect leads the multilayer growth to be dominated by the chaotropic anions in the mixed electrolyte solutions. This is why the theoretical values of $-\Delta f$ are always higher than the experimental values induced by the growth of multilayer in the mixed electrolyte solutions.

Indeed, PEMs not only can be fabricated in water but also can be constructed in organic solvents and water-organic solvent mixtures [22]. Organic solvents and water-organic solvent mixtures usually have a smaller dielectric constant than water. Thus, more ion pairs will be formed compared with that in aqueous solutions [47]. Besides, the solvation of ions in water-organic solvent mixtures should be different from that in water because the solvent molecules can form complexes in the mixtures [48]. Therefore, it is expected that the specific ion effect in organic

4.3 Specific Ion Effect on the Growth of PEMs

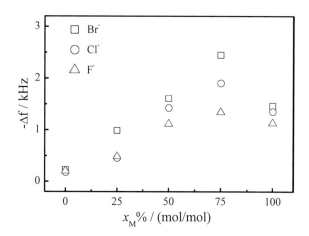

Fig. 4.12 Shift in frequency ($-\Delta f$) for the 8-bilayer PAMPS/PDDA multilayer as a function of the molar fraction of methanol (x_M) in the presence of different anions, where the overtone number $n = 3$ and the salt concentration is fixed at 2.0 mM. Reprinted with permission from Ref. [23]. Copyright 2013 American Chemical Society

solvents and water-organic solvent mixtures might be different from that in water. Figure 4.12 shows the change in $-\Delta f$ for eight bilayers of poly(sodium 2-acrylamido-2-methylpropanesulfonate)/poly(diallyldimethylammonium chloride) (PAMPS/PDDA) multilayer as a function of the molar fraction of methanol (x_M) for the different anions. For all the cases, $-\Delta f$ increases with the x_M from 0 to 75 %, followed by a decrease of $-\Delta f$ with the further increase of x_M from 75 to 100 %. It is interesting that the specific anion effect is observed in the growth of PAMPS/PDDA multilayer at the salt concentration as low as 2.0 mM. Namely, $-\Delta f$ increases following the series $F^- < Cl^- < Br^-$ at the same x_M. More interestingly, the specific anion effect is gradually amplified with the increasing x_M from 0 to 75 % but is weakened again as x_M increases further from 75 to 100 %. It is reported that the dielectric constant (ε) of water–methanol mixtures gradually decreases from 78.5 to 32.7 as x_M increases from 0 to 100 % at 25 °C [49]. At x_M of 0 %, no obvious specific anion effect is observed. This is understandable because l_D is of ~ 7 nm in the presence of 2.0 mM monovalent salt in water and the short-range ion-specific interactions are masked by the long-range nonspecific electrostatic interactions. l_D decreases with the decreasing ε upon the addition of methanol, so that the anion-specific interactions and the specific anion effect should become more obvious with the increasing x_M. This may explain the occurrence of specific anion effect in methanol and water–methanol mixtures but cannot explain why the strongest specific anion effect occurs at the x_M of 75 %.

The concept of matching water affinities could be extended to methanol and water–methanol mixtures, i.e., matching solvent affinities. For the same solvent, the ionic solvation is mainly determined by the charge-dipole interactions between ions and solvent molecules. Thus, it is reasonable to expect that two oppositely charged ions should have similar strengths of solvation in methanol and in water–methanol mixtures if they have similar strengths of hydration in water. In other words, two oppositely charged ions should have stronger interactions in methanol and in water–methanol mixtures if they interact more strongly in water. In aqueous solutions, the strength of interactions between the weakly hydrated ammonium groups on PDDA chains and the anions increases following the order $F^- < Cl^- < Br^-$ because the extent of hydration of anions decreases from F^- to Br^- [50]. Therefore, the strength of interactions between the ammonium groups and the anions is also expected to increase following the order $F^- < Cl^- < Br^-$ in methanol and in water–methanol mixtures. That is, the effectiveness of anions to screen the charges on PDDA chains increases following the series $F^- < Cl^- < Br^-$, and PDDA chains would adopt a more coiled conformation as the anions change from F^- to Br^-. A more coiled conformation is more favorable for the multilayer growth, so that $-\Delta f$ increases following the order $F^- < Cl^- < Br^-$ at the same x_M. In short, the anion-solvent interactions and the resulted counterion-charged group interactions may be responsible for the occurrence of specific anion effect.

Previous Raman spectra studies showed that water and methanol molecules are able to form complexes with a stoichiometry of $(H_2O)_2(CH_3OH)_5$ [51]. That is, when x_M is less than 75 %, the methanol molecules might not be sufficient to complex with all the water molecules, and more complexes are formed in the solvent mixtures with the increasing x_M. Further increasing x_M from 75 to 100 % leads the concentration of complexes to decrease because water molecules may not be sufficient to complex with all the methanol molecules. Thus, the complexes might have the highest concentration at x_M of ~75 % where water and methanol molecules all form complexes. The strength of charge-dipole interactions between the anions and the solvent complexes should decrease following the order $F^- > Cl^- > Br^-$ in the water–methanol mixtures as the anionic surface charge density decreases from F^- to Br^-. Relatively strong interactions between the anions and the solvent complexes lead to weaker interactions between the anions and the ammonium groups, whereas relatively weak interactions between the anions and the solvent complexes cause stronger interactions between the anions and the ammonium groups. As mentioned above, the concentration of solvent complexes increases with x_M from 0 to 75 %. The increasing concentration of water/methanol complexes would amplify the difference in the interactions between anions and solvent complexes, thereby enlarging the difference in the interactions between anions and ammonium groups, giving rise to an amplification of the specific anion effect. In contrast, as x_M increases from 75 to 100 %, the decrease of concentration of solvent complexes would weaken the specific anion effect. Thus, the change of relative proportion of the complexes in the water–methanol mixtures may be responsible for the amplification or weakening of the specific anion effect.

The specific cation effect is usually much weaker than that of anions. However, the cation specificity can be observed in Fig. 4.13; namely, $-\Delta f$ increases following the series $Li^+ < Na^+ < K^+$ at the same x_M. Moreover, the specific cation effect is amplified as x_M increases from 0 to 75 % but is weakened again as x_M increases further from 75 to 100 %. This result is similar to the observation in

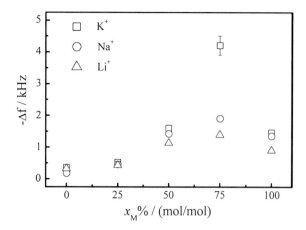

Fig. 4.13 Shift in frequency ($-\Delta f$) for the 8-bilayer PAMPS/PDDA multilayer as a function of the molar fraction of methanol (x_M) in the presence of different cations, where the overtone number $n = 3$ and the salt concentration is fixed at 2.0 mM. Reprinted with permission from Ref. [23]. Copyright 2013 American Chemical Society

Fig. 4.12 and can also be explained by the concept of matching solvent affinities in the water–methanol mixtures.

4.4 Effects of Chain Rigidity and Architecture on the Growth of PEMs

In general, the conformation of polyelectrolyte is determined by the total chain persistence length (l_p) which represents the effective rigidity of the polyelectrolyte chain. l_p is the sum of l_0 and the electrostatic persistence length (l_e) [52]:

$$l_p = l_0 + l_e = l_0 + \frac{l_B l_D^2 \sigma^2}{4} \quad (4.2)$$

where l_B and σ are the Bjerrum length and the linear charge density, respectively. l_0 corresponding to the rigidity of an uncharged chain is independent of the salt concentration, while l_e arising from the electrostatic repulsion between identically charged groups from the same chain depends on the external salt concentration [40]. In the growth of PEMs, increasing salt concentration can screen not only the electrostatic attraction between oppositely charged chains from the neighboring layers but also the electrostatic repulsion between the groups with like charges along the same chain. The former is unfavorable for the multilayer growth due to the decrease of chain interpenetration [53]. In contrast, the latter would give rise to a more coiled conformation of the chains on the surface and thus favors the multilayer growth via increasing the extents of surface charge overcompensation and chain interpenetration [8, 41]. Consequently, the effect of salt concentration on the multilayer growth should be determined by the delicate balance between such two opposite effects.

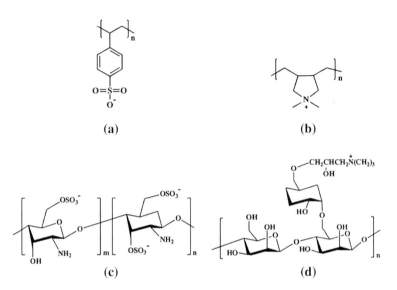

Fig. 4.14 Chemical structures of four types of polyelectrolytes. **a** poly(sodium 4-styrenesulfonate) (PSS), **b** poly(diallyldimethylammonium chloride) (PDDA), **c** sulfated chitosan (SC), **d** cationic guar gum (CGG). Reprinted with permission from Ref. [28]. Copyright 2012 American Chemical Society

The chemical structures of PSS, PDDA, sulfated chitosan (SC), and cationic guar gum (CGG) are shown in Fig. 4.14. The l_0 of PSS, PDDA, SC, and CGG are ~ 0.9, ~ 2.7, ~ 10.0, and ~ 10.0 nm, respectively [54–57]. Thus, PSS and PDDA are considered as flexible polyelectrolytes, whereas SC and CGG are classified as semiflexible polyelectrolytes. As can be seen from Fig. 4.5, the increase of salt concentration favors the growth of PSS/PDDA multilayer; namely, both $-\Delta f$ and ΔD increase with C_{NaCl} for the same layer number, indicating that the effect of weakening of electrostatic repulsion on the multilayer growth dominates over that of the weakening of electrostatic attraction. Thus, the growth of PEMs formed by two flexible polyelectrolytes is dominated by the weakening of electrostatic repulsion between the identically charged groups with the salt concentration.

Figure 4.15 shows the layer number dependence of shifts in $-\Delta f$ and ΔD for the growth of SC/CGG multilayer as a function of C_{NaCl}. In Fig. 4.15a, as C_{NaCl} increases, $-\Delta f$ exhibits two different regimes. For the same layer number, $-\Delta f$ increases with salt concentration in the range of $C_{NaCl} < 0.1$ M, and then $-\Delta f$ decreases with the increasing salt concentration at $C_{NaCl} > 0.1$ M. Likewise, ΔD exhibits a similar result. As C_{NaCl} increases, ΔD increases for the same layer number at $C_{NaCl} < 0.1$ M. When C_{NaCl} is above 0.1 M, ΔD decreases with the increase of salt concentration for the same layer number. Obviously, the salt effect on the growth of SC/CGG multilayer is quite different from that of PSS/PDDA multilayer. Here, the increase of C_{NaCl} favors the growth of SC/CGG multilayer at

4.4 Effects of Chain Rigidity and Architecture on the Growth of PEMs   63

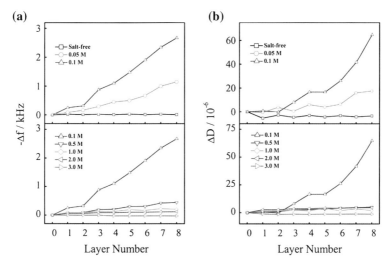

Fig. 4.15 **a** The layer number dependence of shift in frequency ($-\Delta f$) for the growth of SC/CGG multilayer as a function of NaCl concentration (C_{NaCl}), **b** The layer number dependence of shift in dissipation (ΔD) for the growth of SC/CGG multilayer as a function of NaCl concentration (C_{NaCl}). Here, the overtone number $n = 3$, and the odd and even layer numbers correspond to the deposition of SC and CGG, respectively. Reprinted with permission from Ref. [28]. Copyright 2012 American Chemical Society

low salt concentrations, but the increasing C_{NaCl} is unfavorable for the growth of SC/CGG multilayer at high salt concentrations.

At low salt concentrations, l_p comes from the contributions of l_0 and l_e. As the salt concentration increases, l_e decreases, leading to a decrease of l_p. Therefore, the polyelectrolyte chains would adopt a more coiled conformation at the polymer/solution interface with the increasing salt concentration due to the weakening of electrostatic repulsion between identically charged groups, which favors the multilayer growth by increasing the extents of surface charge overcompensation and chain interpenetration. Meanwhile, the strength of electrostatic attraction between oppositely charged chains from the neighboring layers decreases with the increasing salt concentration, which is unfavorable for the multilayer growth. Clearly, at low salt concentrations, the increase of $-\Delta f$ and ΔD with C_{NaCl} for the same layer number indicates that the growth of SC/CGG multilayer is dominated by the weakening of electrostatic repulsion between identically charged groups with the salt concentration.

When C_{NaCl} is above 0.1 M, the l_e of SC and CGG are, respectively, less than 0.26 and 0.05 nm calculated from Eq. 4.2; that is, l_e is much smaller than l_0 for these two semiflexible polyelectrolytes. Consequently, as C_{NaCl} increases, the contribution of l_e to l_p can be neglected and l_p would keep almost constant with C_{NaCl}. As a result, the chain conformation and the extent of surface charge overcompensation should only slightly change with C_{NaCl} even though the electrostatic repulsion is gradually screened by the added salts. On the other hand, the electrostatic attraction between the neighboring layers is also screened with the increase of C_{NaCl}, which is unfavorable for the multilayer growth due to the decrease of chain interpenetration. The fact that $-\Delta f$ and ΔD decrease with salt concentration for the same layer number indicates that the effect of weakening of electrostatic attraction on the multilayer growth dominates over that of the weakening of electrostatic repulsion. That is, as C_{NaCl} increases, the growth of SC/CGG multilayer is dominated by the weakening of electrostatic attraction between the neighboring layers at the high salt concentrations.

In comparison with linear–linear PEMs, multilayers formed by star polyelectrolytes usually exhibit some unique properties [30]. Such unique properties of star–star PEMs should be attributed to the topological structure of star polyelectrolytes and the resulted distinct behavior of chain interpenetration between the layers. Star-shaped polyelectrolytes usually have a more compact structure, giving rise to a more limited interpenetration in comparison with that of linear chains [30]. Thus, it is anticipated that the behavior of chain interpenetration in the growth of star–star PEMs should be different from that in the growth of linear–linear PEMs. As the arm number of star polyelectrolytes changes, the resulted steric effect can also influence the chain interpenetration. Actually, QCM-D not only can tell the extent of chain interpenetration, but also can provide which kind of polyelectrolyte will penetrate into the oppositely charged layer.

In Fig. 4.16, linear poly(acrylic acid) (PAA) and star-shaped PDEM with different arm numbers but similar arm lengths are used to fabricate the PEMs. The number combinations (e.g., 2-2) in the figure legend represent the different PAA-PDEM pairs. The first and second numbers denote the arm number of PAA and PDEM, respectively. The arm number of PAA is fixed at two (linear chain), and the arm number of PDEM is gradually increased from 2 to 6. In the case of the 2-2 pair, ΔD increases in an oscillatory manner with the layer number, indicating the alternating deposition of PAA and PDEM on the resonator surface. The increase in ΔD for the odd layer number indicates that the deposited PAA chains form a swollen layer on the surface, whereas the decrease in ΔD for the even layer number implies that the adsorbed PDEM chains penetrate into the predeposited PAA layer and form a relatively dense layer through polyelectrolyte complexation. Consequently, the oscillating changes in ΔD with the layer number indicate the alternating swelling and shrinking of the outermost layer of the multilayer due to the chain interpenetration. Obviously, in the case of 2-2 pair, the PDEM chains penetrate into the PAA layer. Likewise, in the case of 2-3 pair, the arm chains of PDEM also penetrate into the PAA layer, as reflected by the fact that ΔD increases and decreases for the odd and even layer numbers, respectively.

However, no obvious oscillations in ΔD with the layer number are observed for the 2-4 pair, indicating that only slight chain interpenetration occurs in this case. This is understandable because the more prominent steric effect created by star PDEM prevents the chain interpenetration as the arm number increases to 4. When the arm number is increased to 6, the steric effect created by star PDEM should be more notable. Nonetheless, the chain interpenetration is observed again in the case

Fig. 4.16 Shifts in dissipation (ΔD) and frequency ($-\Delta f$) as a function of layer number for the growth of PAA/PDEM multilayer, where the overtone number $n = 3$ and pH is fixed at ~ 5.3. The odd and even layer numbers correspond to the deposition of PAA and PDEM, respectively. The arm number of PAA is fixed at 2 for all the pairs and the arm number of PDEM is gradually increased from 2 to 6. Reprinted with permission from Ref. [31]. Copyright 2012 American Chemical Society

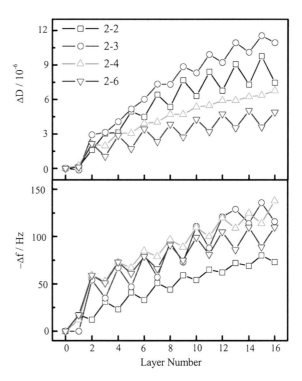

of 2-6 pair. More interestingly, the increase in ΔD for the even layer number and the decrease in ΔD for the odd layer number imply that the PAA chains penetrate into the PDEM layer, which is in contrast to the behavior for the 2-2 and 2-3 pairs. Thus, one can conclude that it is more difficult for star PDEM to penetrate into the PAA layer as the arm number of PDEM increases because of the increasing steric effect. The 2-4 pair is a critical case at which the behavior of chain interpenetration changes dramatically; namely, the arm chains of PDEM penetrate into the PAA layer when the arm number of PDEM is less than 4, whereas the PAA chains penetrate into the PDEM layer when the arm number of PDEM is greater than 4.

On the other hand, the gradual increase of $-\Delta f$ with the layer number further indicates the sequential deposition of polyelectrolytes on the surface. Basically, $-\Delta f$ relates to the mass change of the PEM, which is determined by the competition between the adsorption of polyelectrolyte chains and the release of trapped water molecules from the PEM during polyelectrolyte complexation. In the cases of 2-2 and 2-3 pairs, $-\Delta f$ increases for the odd layer number, indicating an increase of the mass of the PEM induced by the adsorption of PAA chains, and the decrease of $-\Delta f$ for the even layer number indicates a decrease of the mass of the PEM due to the release of trapped water molecules during the penetration of PDEM arm chains into the PAA layer. In other words, the mass change of PEM for the odd and even layer numbers is dominated by the adsorption of PAA chains and the release of trapped water molecules, respectively. The oscillations in $-\Delta f$ imply the alternating increase and decrease of the mass of the PEM during LbL deposition. In the case of 2-6 pair, the increase of $-\Delta f$ for the even layer number indicates an increase of the mass of the PEM because of the deposition of PDEM, and the decrease of $-\Delta f$ for the odd layer number implies a decrease of the mass of the PEM due to the release of trapped water molecules during the penetration of PAA chains into the PDEM layer. In the case of 2-4 pair, the small oscillations in $-\Delta f$ with the layer number might indicate small oscillating changes in the mass of the PEM during the interfacial complexation between PAA and PDEM. In addition, as the arm number of PAA increases, the growth of PAA/PDEM multilayer will exhibit a different behavior of chain interpenetration, that is, the chain interpenetration is also dependent on the chain architecture of PAA [31].

4.5 Conclusion

The influences of external conditions and intrinsic chain properties on the growth of PEMs have been studied by use of QCM-D. QCM-D can provide not only the changes in mass, thickness, and structure of the PEMs but also the changes in chain conformation and interpenetration during the multilayer growth. The influences of temperature, pH, and salt concentration on the growth of PSS/PDEM multilayer are dominated by the chain interpenetration. The growth of PSS/PDDA multilayer is dominated by chain conformation and chain interpenetration at $C_{NaCl} < 1.0$ M and $C_{NaCl} \geq 1.0$ M, respectively. The specific ion effect on the growth of multilayer is controlled by the specific counterion-charged group interaction, which strongly influences the chain conformation and interpenetration. Regarding the effect of chain rigidity on the growth of PEMs, the multilayer growth is determined by the delicate balance between the weakening of electrostatic repulsion and the decrease of electrostatic attraction. In the growth of PEMs formed by polyelectrolytes with different chain architectures, the chain interpenetration is mainly governed by the steric effect created by the arm chains.

References

1. Decher G (1997) Fuzzy nanoassemblies: toward layered polymeric multicomposites. Science 277:1232–1237
2. Caruso F, Caruso RA, Mohwald H (1998) Nanoengineering of inorganic and hybrid hollow spheres by colloidal templating. Science 282:1111–1114
3. Schonhoff M (2003) Self-assembled polyelectrolyte multilayers. Curr Opin Colloid Interface Sci 8:86–95
4. Eckle M, Decher G (2001) Tuning the performance of layer-by-layer assembled organic light emitting diodes by controlling the position of isolating clay barrier sheets. Nano Lett 1:45–49
5. Hiller J, Mendelsohn JD, Rubner MF (2002) Reversibly erasable nanoporous anti-reflection coatings from polyelectrolyte multilayers. Nat Mater 1:59–63

References

6. Lichter JA, Van Vliet KJ, Rubner MF (2009) Design of antibacterial surfaces and interfaces: Polyelectrolyte multilayers as a multifunctional platform. Macromolecules 42:8573–8586
7. Palama IE, Leporatti S, de Luca E, Di Renzo N, Maffia M, Gambacorti-Passerini C, Rinaldi R, Gigli G, Cingolani R, Coluccia AML (2010) Imatinib-loaded polyelectrolyte microcapsules for sustained targeting of BCR-ABL(+) leukemia stem cells. Nanomedicine 5:419–431
8. Liu GM, Zhao JP, Sun QY, Zhang GZ (2008) Role of chain interpenetration in layer-by-layer deposition of polyelectrolytes. J Phys Chem B 112:3333–3338
9. Gopinadhan M, Ivanova O, Ahrens H, Gunther JU, Steitz R, Helm CA (2007) The influence of secondary interactions during the formation of polyelectrolyte multilayers: Layer thickness, bound water and layer interpenetration. J Phys Chem B 111:8426–8434
10. Tan HL, McMurdo MJ, Pan GQ, Van Patten PG (2003) Temperature dependence of polyelectrolyte multilayer assembly. Langmuir 19:9311–9314
11. Yoo D, Shiratori SS, Rubner MF (1998) Controlling bilayer composition and surface wettability of sequentially adsorbed multilayers of weak polyelectrolytes. Macromolecules 31:4309–4318
12. Shiratori SS, Rubner MF (2000) pH-dependent thickness behavior of sequentially adsorbed layers of weak polyelectrolytes. Macromolecules 33:4213–4219
13. Schoeler B, Poptoschev E, Caruso F (2003) Growth of multilayer films of fixed and variable charge density polyelectrolytes: effect of mutual charge and secondary interactions. Macromolecules 36:5258–5264
14. Ladam G, Schaad P, Voegel JC, Schaaf P, Decher G, Cuisinier F (2000) In situ determination of the structural properties of initially deposited polyelectrolyte multilayers. Langmuir 16:1249–1255
15. Huang SCJ, Artyukhin AB, Wang YM, Ju JW, Stroeve P, Noy A (2005) Persistence length control of the polyelectrolyte layer-by-layer self-assembly on carbon nanotubes. J Am Chem Soc 127:14176–14177
16. Liu GM, Zou SR, Fu L, Zhang GZ (2008) Roles of chain conformation and interpenetration in the growth of a polyelectrolyte multilayer. J Phys Chem B 112:4167–4171
17. Salomaki M, Tervasmaki P, Areva S, Kankare J (2004) The Hofmeister anion effect and the growth of polyelectrolyte multilayers. Langmuir 20:3679–3683
18. Wong JE, Zastrow H, Jaeger W, von Klitzing R (2009) Specific ion versus electrostatic effects on the construction of polyelectrolyte multilayers. Langmuir 25:14061–14070
19. Liu GM, Hou Y, Xiao XA, Zhang GZ (2010) Specific anion effects on the growth of a polyelectrolyte multilayer in single and mixed electrolyte solutions investigated with quartz crystal microbalance. J Phys Chem B 114:9987–9993
20. Dubas ST, Schlenoff JB (1999) Factors controlling the growth of polyelectrolyte multilayers. Macromolecules 32:8153–8160
21. Poptoshev E, Schoeler B, Caruso F (2004) Influence of solvent quality on the growth of polyelectrolyte multilayers. Langmuir 20:829–834
22. Zhang P, Qian JW, An QF, Du BY, Liu XQ, Zhao Q (2008) Influences of solution property and charge density on the self-assembly behavior of water-insoluble polyelectrolyte sulfonated poly(sulphone) sodium salts. Langmuir 24:2110–2117
23. Long YC, Wang T, Liu LD, Liu GM, Zhang GZ (2013) Ion specificity at a low salt concentration in water-methanol mixtures exemplified by a growth of polyelectrolyte multilayer. Langmuir 29:3645–3653
24. Sui ZJ, Salloum D, Schlenoff JB (2003) Effect of molecular weight on the construction of polyelectrolyte multilayers: stripping versus sticking. Langmuir 19:2491–2495
25. Porcel C, Lavalle P, Decher G, Senger B, Voegel JC, Schaaf P (2007) Influence of the polyelectrolyte molecular weight on exponentially growing multilayer films in the linear regime. Langmuir 23:1898–1904
26. Johansson E, Lundstrom L, Norgren M, Wagberg L (2009) Adsorption behavior and adhesive properties of biopolyelectrolyte multilayers formed from cationic and anionic starch. Biomacromolecules 10:1768–1776

27. Mjahed H, Cado G, Boulmedais F, Senger B, Schaaf P, Ball V, Voegel JC (2011) Restructuring of exponentially growing polyelectrolyte multilayer films induced by salt concentration variations after film deposition. J Mater Chem 21:8416–8421
28. Wu B, Li CL, Yang HY, Liu GM, Zhang GZ (2012) Formation of polyelectrolyte multilayers by flexible and semiflexible chains. J Phys Chem B 116:3106–3114
29. Kim BS, Gao HF, Argun AA, Matyjaszewski K, Hammond PT (2009) All-star polymer multilayers as pH-responsive nanofilms. Macromolecules 42:368–375
30. Choi I, Suntivich R, Pamper FA, Synatschke CV, Muller AHE, Tsukruk VV (2011) pH-controlled exponential and linear growing modes of layer-by-layer assemblies of star polyelectrolytes. J Am Chem Soc 133:9592–9606
31. Chen FG, Liu GM, Zhang GZ (2012) Formation of multilayers by star polyelectrolytes: effect of number of arms on chain interpenetration. J Phys Chem B 116:10941–10950
32. Picart C, Mutterer J, Richert L, Luo Y, Prestwich GD, Schaaf P, Voegel JC, Lavalle P (2002) Molecular basis for the explanation of the exponential growth of polyelectrolyte multilayers. Proc Natl Acad Sci 99:12531–12535
33. Laugel N, Betscha C, Winterhalter M, Voegel JC, Schaaf P, Ball V (2006) Relationship between the growth regime of polyelectrolyte multilayers and the polyanion/polycation complexation enthalpy. J Phys Chem B 110:19443–19449
34. Salomaki M, Vinokurov IA, Kankare J (2005) Effect of temperature on the buildup of polyelectrolyte multilayers. Langmuir 21:11232–11240
35. Cho J, Quinn JF, Caruso F (2004) Fabrication of polyelectrolyte multilayer films comprising nanoblended layers. J Am Chem Soc 126:2270–2271
36. Lavalle P, Picart C, Mutterer J, Gergely C, Reiss H, Voegel JC, Senger B, Schaaf P (2004) Modeling the buildup of polyelectrolyte multilayer films having exponential growth. J Phys Chem B 108:635–648
37. Schoeler B, Kumaraswamy G, Caruso F (2002) Investigation of the influence of polyelectrolyte charge density on the growth of multilayer thin films prepared by the layer-by-layer technique. Macromolecules 35:889–897
38. Lee AS, Gast AP, Butun V, Armes SP (1999) Characterizing the structure of pH dependent polyelectrolyte block copolymer micelles. Macromolecules 32:4302–4310
39. Losche M, Schmitt J, Decher G, Bouwman WG, Kjaer K (1998) Detailed structure of molecularly thin polyelectrolyte multilayer films on solid substrates as revealed by neutron reflectometry. Macromolecules 31:8893–8906
40. McAloney RA, Sinyor M, Dudnik V, Goh MC (2001) Atomic force microscopy studies of salt effects on polyelectrolyte multilayer film morphology. Langmuir 17:6655–6663
41. Schlenoff JB, Dubas ST (2001) Mechanism of polyelectrolyte multilayer growth: charge overcompensation and distribution. Macromolecules 34:592–598
42. Marcus Y (2009) Effect of ions on the structure of water: structure making and breaking. Chem Rev 109:1346–1370
43. Collins KD (2004) Ions from the Hofmeister series and osmolytes: effects on proteins in solution and in the crystallization process. Methods 34:300–311
44. Collins KD (2006) Ion hydration: Implications for cellular function, polyelectrolytes, and protein crystallization. Biophys Chem 119:271–281
45. Parsons DF, Ninham BW (2010) Importance of accurate dynamic polarizabilities for the ionic dispersion interactions of alkali halides. Langmuir 26:1816–1823
46. Lima ERA, Bostrom M, Horinek D, Biscaia EC, Kunz W, Tavares FW (2008) Co-ion and ion competition effects: Ion distributions close to a hydrophobic solid surface in mixed electrolyte solutions. Langmuir 24:3944–3948
47. Ibuki K, Nakahara M (1987) Effect of dielectric friction on the viscosity B-coefficient for electrolyte in methanol-water mixture. J Chem Phys 86:5734–5738
48. Dixit S, Crain J, Poon WCK, Finney JL, Soper AK (2002) Molecular segregation observed in a concentrated alcohol–water solution. Nature 416:829–832
49. Albright PS, Gosting LJ (1946) Dielectric constants of the methanol water system from 5-degrees to 55-degrees. J Am Chem Soc 68:1061–1063

50. Vlachy N, Jagoda-Cwiklik B, Vacha R, Touraud D, Jungwirth P, Kunz W (2009) Hofmeister series and specific interactions of charged headgroups with aqueous ions. Adv Colloid Interface 146:42–47
51. Dixit S, Poon WCK, Crain J (2000) Hydration of methanol in aqueous solutions: a Raman spectroscopic study. J Phys Condens Mater 12:L323–L328
52. Dobrynin AV (2005) Electrostatic persistence length of semiflexible and flexible polyelectrolytes. Macromolecules 38:9304–9314
53. Von Klitzing R, Wong JE, Jaeger W, Steitz R (2004) Short range interactions in polyelectrolyte multilayers. Curr Opin Colloid Interface Sci 9:158–162
54. Brulet A, Boue F, Cotton JP (1996) About the experimental determination of the persistence length of wormlike chains of polystyrene. J Phys II 6:885–891
55. Mattison KW, Dubin PL, Brittain IJ (1998) Complex formation between bovine serum albumin and strong polyelectrolytes: effect of polymer charge density. J Phys Chem B 102:3830–3836
56. Rinaudo M (2006) Non-covalent interactions in polysaccharide systems. Macromol Biosci 6:590–610
57. Morris GA, Patel TR, Picout DR, Ross-Murphy SB, Ortega A, de la Torre JG, Harding SE (2008) Global hydrodynamic analysis of the molecular flexibility of galactomannans. Carbohydr Polym 72:356–360

Chapter 5
Interactions between Polymers and Phospholipid Membranes

Abstract The adsorption of poly(ethylene glycol) (PEG) with different hydrophobic end groups onto phospholipid membranes has been investigated by using a QCM-D in real time. On a SiO_2-coated resonator surface, the adsorption of lipid vesicles results in a solid-supported lipid bilayer (s-SLB). PEG chains with strongly hydrophobic end groups can insert in the bilayer, whereas PEG chains with weakly hydrophobic end groups do not interact with the s-SLB. On the other hand, the adsorbed vesicles are intact on a gold surface. When the end group of PEG chains is not hydrophobic enough, PEG chains interact weakly with the vesicles so that they only have slight effects on the vesicle stability. However, PEG chains with strongly hydrophobic end groups can lead to a vesicle-to-bilayer transition due to the insertion of the chains in the lipid membrane. In addition, PEG can more readily induce the rupture of vesicles at a higher polymer concentration due to the combined effect of hydrophobic interaction and osmotic pressure.

Keywords Adsorption · Bilayer · Hydrophobic interaction · Insertion · Lipid membrane · Osmotic pressure · Poly(ethylene glycol) · Vesicle

5.1 Introduction

Cell membranes are complex dynamical structures primarily consisting of a bilayer in which two layers of phospholipid molecules are arranged in a way that the hydrophilic heads shield the hydrophobic lipid tails from the water [1–3]. As a permeability barrier, the membrane can protect the cell from the environment and maintain membrane protein stability and function [1–3]. Many cellular processes such as endocytosis, exocytosis, fertilization, signal transduction, viral infection, intracellular transport, and cell aggregation are mediated by such a biomembrane [4]. For example, membrane fusion, a ubiquitous life process, is generally controlled by the fusion proteins inserted in the core of cell membrane [5, 6]. Probably because of the complex nature of the cell membranes, direct investigations on biological membranes are extremely difficult to address. Therefore, model membranes are essential for studying the membrane related phenomena. The most commonly used model system is a lipid membrane decorated with polymers [7–9].

In principle, polymers that will be inserted in the lipid membrane should first adsorb and bind on the membrane. One fundamental problem is the effect of molecular interaction on the insertion. Considering that the membrane is composed of a number of amphiphilic phospholipid molecules, the fundamental issue related to the binding of polymer chains onto the membrane surface is the hydrophobic interaction between the anchoring groups and the hydrophobic tails in the membrane. Actually, many life processes ranging from membrane fusion to viral infection are mediated by the hydrophobic interactions between the proteins and the biomembranes [10–12]. It is generally accepted that the mismatch between the hydrophobic thickness of lipid bilayer and the hydrocarbon length of the integral membrane proteins plays an important role in lipid-protein interaction [10]. Without a good match, the proteins will not incorporate into biomembranes [11]. To reduce the hydrophobic mismatch, either the structure of membrane protein or the lipid membrane has to be altered [12]. In this chapter, QCM-D is used to investigate the adsorption of poly(ethylene glycol) (PEG) with different hydrophobic end groups onto the phospholipid membranes to clarify the hydrophobic interactions between the end group of polymer chains and the core of lipid membranes.

5.2 Role of Hydrophobic Interaction in the Adsorption of PEG on Lipid Membrane Surface

Figure 5.1 shows the typical shifts in Δf and ΔD as a function of time for the adsorption of lipid vesicles onto the resonator surfaces. In Fig. 5.1a, Δf decreases and ΔD increases rapidly in the initial stage and then gradually level off, indicating the saturation of lipid vesicles on the gold-coated surface. The monotonic changes in Δf and ΔD demonstrate that the lipid vesicles absorbed on the gold surface are intact. In contrast, the adsorption of lipid vesicles onto the SiO_2-coated resonator surface looks quite different (Fig. 5.1b). Δf first decreases and then increases with a minimum, while ΔD exhibits an opposite behavior with a maximum in the adsorption isotherm. The changes of Δf and ΔD in Fig. 5.1b are indicative of a vesicle-to-bilayer transition [13, 14]. Specifically, the initial decrease in Δf and increase in ΔD indicate that the intact vesicles are adsorbed onto the resonator surface. The following increase in Δf and decrease in ΔD reflect that the vesicles rupture and fuse into a bilayer. In other words, the adsorbed lipid vesicles form a continuously solid-supported lipid bilayer (s-SLB) on the SiO_2-coated resonator surface.

5.2 Role of Hydrophobic Interaction

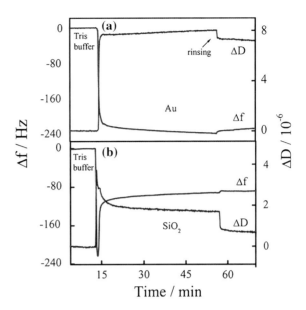

Fig. 5.1 Shifts in frequency (Δf) and dissipation (ΔD) as a function of time for the adsorption of lipid vesicles on the resonator surfaces, where the overtone number n = 3 and the concentration of lipid vesicle is fixed at 3.0 mg/mL. a The gold-coated resonator surface, and b The SiO_2-coated resonator surface. Reprinted with the permission from Ref. [15]. Copyright 2010 American Chemical Society

Figure 5.2 shows the shifts in Δf and ΔD as a function of time for the adsorption of PEG-OH and PEG-CH_3 onto the s-SLB surface. The addition of PEG-OH solution into QCM cell does not lead to any significant change in either Δf or ΔD, suggesting that PEG-OH chains only slightly adsorb on the bilayer surface. That is, the PEG-OH chains only weakly interact with the bilayer surface. Both Δf and ΔD return to the original point after rinsing, indicating that no PEG-OH chains are stably adsorbed on the s-SLB surface. The adsorption of PEG-CH_3 is analogous to that of PEG-OH, though the methyl in the former is more hydrophobic than the hydroxyl in the latter. Therefore, both PEG-OH and PEG-CH_3 chains only weakly interact with the membrane.

When the PEG chains are ended with a much more hydrophobic group (i.e., -$C_{18}H_{37}$), the adsorption of PEG chains onto the s-SLB surface leads to a different behavior of shifts in Δf and ΔD with time (Fig. 5.3). The critical micelle concentration (CMC) of PEG-$C_{18}H_{37}$ is ~0.1 mg/mL [17], so PEG-$C_{18}H_{37}$ chains exist as individuals at the concentration of 0.05 mg/mL. After PEG-$C_{18}H_{37}$ solution is introduced at C of 0.05 mg/mL, Δf sharply decreases, indicating the insertion of PEG-$C_{18}H_{37}$ chains in the lipid bilayer. Also, the inserted chains and the incoming chains may form aggregates on the membrane surface even at the polymer concentration below CMC due to the attractions between the chains. Subsequently, Δf slowly increases, implying that the trapped water molecules in the polymer aggregates are slowly released out due to the rearrangement of polymer chains on the membrane surface. On the other hand, the sharp increase of ΔD in the initial stage further indicates that PEG-$C_{18}H_{37}$ chains insert into the bilayer and they form random aggregates with the incoming chains. The subsequent decrease in ΔD suggests that the adsorbed chains gradually pack more

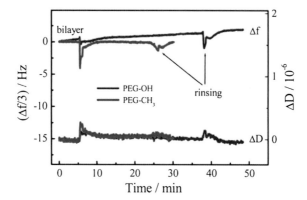

Fig. 5.2 Shifts in frequency (Δf) and dissipation (ΔD) as a function of time for the adsorption of PEG-OH and PEG-CH$_3$ onto the solid-supported lipid bilayer surface, where the overtone number n = 3 and the polymer concentration is fixed at 0.05 mg/mL. Reprinted with the permission from Ref. [16]. Copyright 2009 American Chemical Society

Fig. 5.3 Shifts in frequency (Δf) and dissipation (ΔD) as a function of time for the adsorption of PEG-C$_{18}$H$_{37}$ onto the solid-supported lipid bilayer surface with the polymer concentration of 0.05 and 1.0 mg/mL, where the overtone number n = 3. Reprinted with the permission from Ref. [16]. Copyright 2009 American Chemical Society

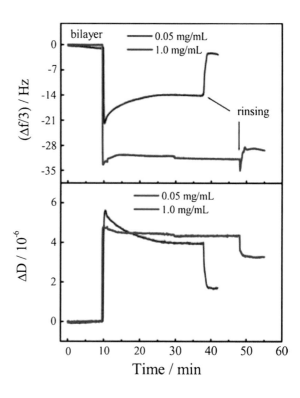

densely upon the rearrangement of the chains. After rinsing, Δf almost returns to the baseline, indicating that most of the polymer aggregates are removed from the outer surface of membrane and only a small amount of polymer chains are incorporated in the bilayer. Thus, the difference in adsorption between PEG-$C_{18}H_{37}$ and PEG-OH or PEG-CH_3 is mainly due to the insertion and aggregation of the chains with different hydrophobic end groups. However, ΔD does not return to the baseline after rinsing, which is because the hydrophilic tails of the incorporated PEG-$C_{18}H_{37}$ chains protruding from the s-SLB surface have a significant influence on ΔD [18]. At C of 1.0 mg/mL, PEG-$C_{18}H_{37}$ chains form micelles in equilibrium with free chains. The initial sharp decrease in Δf and increase in ΔD indicate the rapid insertion of the free chains. Then, Δf and ΔD keep almost constant with time, implying no obvious rearrangement of polymer chains on the membrane surface. The rinsing with buffer only leads to small changes in Δf and ΔD, indicating that the incorporated chains are stably attached on the s-SLB surface. This might be because the higher polymer concentration outside the bilayer with a higher osmotic pressure causes a stronger adsorption of PEG-$C_{18}H_{37}$ chains on the membrane surface.

Figure 5.4 shows the shifts in Δf and ΔD as a function of time for the adsorption of PEG-OH and PEG-CH_3 onto the layer formed by lipid vesicles at a concentration of 0.05 mg/mL. The introduction of PEG-OH only leads to a slight change in Δf, indicating that PEG-OH chains only slightly adsorb on lipid vesicle surface. The relatively large shift in ΔD might be attributed to the formation of loops or tails of a few PEG-OH chains on the vesicle surface, which has a marked effect on

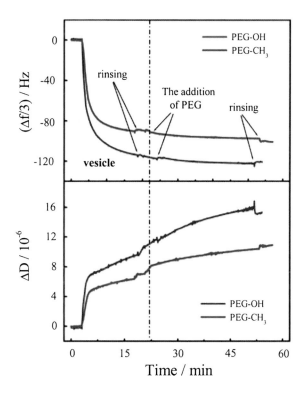

Fig. 5.4 Shifts in frequency (Δf) and dissipation (ΔD) as a function of time for the adsorption of PEG-OH and PEG-CH_3 onto the layer formed by lipid vesicles, where the overtone number $n = 3$ and the polymer concentration is fixed at 0.05 mg/mL. Reprinted with the permission from Ref. [16]. Copyright 2009 American Chemical Society

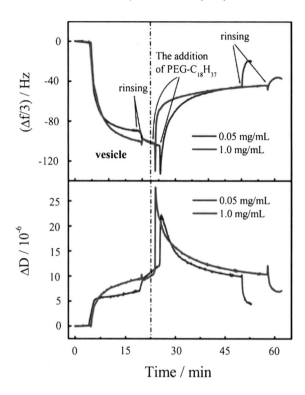

Fig. 5.5 Shifts in frequency (Δf) and dissipation (ΔD) as a function of time for the adsorption of PEG-$C_{18}H_{37}$ onto the layer formed by lipid vesicles with the polymer concentration of 0.05 and 1.0 mg/mL, where the overtone number n = 3. Reprinted with the permission from Ref. [16]. Copyright 2009 American Chemical Society

the energy dissipation. The adsorption of PEG-CH_3 chains onto the vesicle surface has a similar phenomenon with that of PEG-OH. Thus, either PEG-OH or PEG-CH_3 chains do not exhibit significant interactions with the vesicle membrane surface, so that no vesicle-to-bilayer transition occurs.

In Fig. 5.5, Δf decreases and ΔD increases sharply in the initial stage after adding PEG-$C_{18}H_{37}$, indicating a rapid binding of PEG-$C_{18}H_{37}$ chains onto the surface of lipid vesicles. The most important event is that Δf and ΔD exhibit a minimum and a maximum, respectively. The increase in Δf and decrease in ΔD after their extrema indicate that the vesicles rupture and fuse into a bilayer accompanied by the release of trapped water molecules during the vesicle-to-bilayer transition. Theoretical studies indicate that the adsorption of vesicles onto a surface is governed by the competition between adhesion energy (F_a) and bending energy (F_b), where $F_a = -WA^*$ and $F_b = (k/2) \oint dA(C_1 + C_2 - C_0)^2$ [19–21]. W and A^* are the effective contact potential and the contact area, respectively, k is the bending rigidity of the membrane, C_1 and C_2 are the two principal curvatures, and C_0 is the spontaneous curvature [19–21]. The former is the energy to deform the shape, which is gained by the vesicle adsorption. The latter is the energy to hold the vesicle shape. When the latter is dominated by the former, a planar bilayer will be resulted via the fusion of vesicles [22]. During the adsorption of PEG-$C_{18}H_{37}$ chains on the membrane surface, W and A^* should hold constant,

5.2 Role of Hydrophobic Interaction

thereby resulting in a constant F_a. Therefore, the rupture of vesicles is likely dominated by the changes in membrane curvature and bending rigidity induced by the adsorption of polymer chains.

5.3 Effect of Length of Hydrocarbon End Group on the Adsorption of PEG on Lipid Membrane Surface

From the discussion above, PEG-CH$_3$ cannot insert in the lipid bilayer, whereas PEG-C$_{18}$H$_{37}$ can insert in the lipid bilayer. That is, the insertion efficiency increases with the length of hydrocarbon end group of the PEG chains. However, the critical value of the length below which the PEG chains cannot insert in the membrane is still unknown. Figure 5.6 shows the shifts in Δf and ΔD as a function of time for the adsorption of PEG chains on the s-SLB surface. Here, the PEG chains with m carbons in the hydrocarbon end group (PEG-OOC(CH$_2$)$_{m-2}$CH$_3$) is designated as C$_m$ in the figure. When the carbon number $m = 11$, the introduction of PEG-C$_{11}$H$_{21}$ solution does not induce any significant change in either Δf or ΔD, suggesting only slight adsorption of PEG-C$_{11}$H$_{21}$ chains on the s-SLB surface.

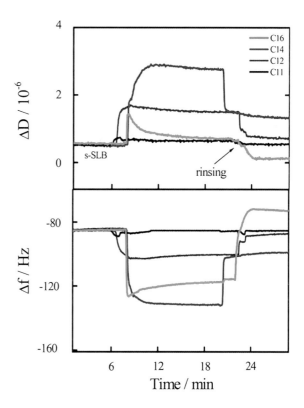

Fig. 5.6 Shifts in frequency (Δf) and dissipation (ΔD) as a function of time for the adsorption of PEG chains on the s-SLB surface, where the overtone number n = 3 and the polymer concentration is fixed at 0.1 mg/mL. PEG with m carbons in the hydrocarbon end group is designated as C$_m$. Reprinted with the permission from Ref. [15]. Copyright 2010 American Chemical Society

Particularly, both Δf and ΔD return to the baseline after rinsing, further indicating that very few PEG-$C_{11}H_{21}$ chains are adsorbed on the s-SLB surface. However, Δf decreases and ΔD increases significantly when the carbon number reaches 12, indicating that the attachment of PEG-$C_{12}H_{23}$ chains on the s-SLB surface. Both Δf and ΔD almost return to the baseline after rinsing, implying that most of the PEG-$C_{12}H_{23}$ chains are removed from the bilayer surface. This is because the hydrophobic interaction between PEG-$C_{12}H_{23}$ and lipid bilayer is not strong enough to suppress the impact of rinsing at $m = 12$. When m increases to 14, Δf and ΔD exhibit large changes even after the rinsing, indicating that the hydrophobic interaction becomes so strong that some PEG-$C_{14}H_{27}$ chains root in the lipid membrane and cannot be removed upon rinsing. When m increases to 16, a more complex behavior is observed. Δf and ΔD exhibit a minimum and a maximum in the adsorption isotherm, respectively. The initial decrease in Δf and increase in ΔD are attributed to the deep insertion of PEG-$C_{16}H_{31}$ chains in the lipid bilayer as well as the association of the inserted chains with the incoming chains on the membrane surface. Subsequently, Δf increases and ΔD decreases slowly, indicating that the trapped water molecules in the PEG aggregates are slowly released out due to the rearrangement of PEG-$C_{16}H_{31}$ chains on the membrane surface, which is similar to the result observed in Fig. 5.3 for the adsorption of PEG-$C_{18}H_{37}$ chains. After rinsing, Δf is higher than the baseline and ΔD is lower than the baseline, indicating that the PEG-$C_{16}H_{31}$ aggregates and some associated phospholipid molecules might be removed from the membrane surface [23].

Considering that the hydrophobic interaction between PEG chains and lipid membrane increases as the length of hydrocarbon end group increases, the adsorption and insertion of the PEG chains on the lipid membrane surface should be driven by the hydrophobic interaction. In other words, the binding of PEG chains on the lipid membrane is controlled by the length of the hydrocarbon end group. Clearly, the critical value of carbon number in the hydrocarbon end group is 12, above which the hydrophobic end group can insert in the lipid bilayer. Additionally, the rinsing would produce a shear force to drive the PEG chains away from the membrane surface if the hydrophobic attraction is not strong enough. This explains why both Δf and ΔD can return to the baseline after rinsing when the hydrocarbon end group is not long enough.

Likewise, the introduction of PEG-$C_{11}H_{21}$ to lipid vesicles does not lead to any obvious change in either Δf or ΔD (Fig. 5.7). However, Δf decreases and ΔD increases markedly when m increases to 12, indicating an adsorption of PEG-$C_{12}H_{23}$ chains onto the lipid vesicle surface. Moreover, Δf and ΔD almost return to the baseline after rinsing, implying that such a adsorption is weak and the adsorbed chains can be removed upon rinsing. Similar phenomena can also be observed at $m = 14$. When m increases to 16, Δf first sharply decreases and then rapidly increases with a minimum, whereas ΔD exhibits an opposite behavior with a maximum. This result is similar to the observation for the adsorption of PEG-$C_{18}H_{37}$ chains (Fig. 5.5) and is indicative of the vesicle-to-bilayer transition. The final value of Δf after rinsing is higher than the baseline, indicating that water

5.3 Effect of Length of Hydrocarbon End

Fig. 5.7 Shifts in frequency (Δf) and dissipation (ΔD) as a function of time for the adsorption of PEG chains onto the layer formed by lipid vesicles, where the overtone number $n = 3$ and the polymer concentration is fixed at 0.1 mg/mL. PEG with m carbons in the hydrocarbon end group is designated as C_m. Reprinted with the permission from Ref. [15]. Copyright 2010 American Chemical Society

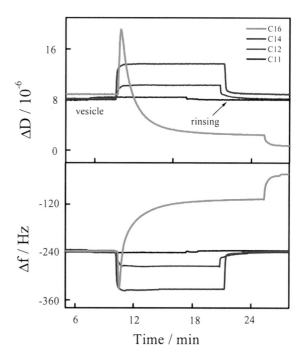

molecules trapped in the vesicles are released out after the vesicle-to-bilayer transition. At the same time, ΔD after rinsing is lower than the baseline, further indicating the transition from soft vesicles to a bilayer.

When the PEG concentration is increased to 2.0 mg/mL, the increasing osmotic pressure of polymer solution to the membrane is favorable for the insertion of the PEG chains. In Fig. 5.8, Δf decreases and ΔD increases after the introduction of PEG-$C_{11}H_{21}$ solution, indicating that the PEG-$C_{11}H_{21}$ chains can adsorb on the vesicle surface at such a high concentration. Both Δf and ΔD return to the baseline after rinsing, implying that the adsorption is weak and the adsorbed chains can be removed by rinsing. When m increases to 12, Δf and ΔD show a minimum and a maximum, respectively, indicating that lipid vesicles rupture and fuse into a bilayer due to the adsorption and insertion of PEG-$C_{12}H_{23}$ chains. This is in contrast with the case at the PEG concentration of 0.1 mg/mL, where PEG-$C_{12}H_{23}$ chains cannot induce the rupture of vesicles (Fig. 5.7). Therefore, the adsorption and insertion of PEG chains are not only related to the hydrophobic interaction but also to the osmotic pressure between the vesicle core and the PEG solution outside the membrane. The osmotic pressure increases with the PEG concentration, which would promote the insertion of PEG chains in the lipid vesicle membrane and accelerate the lipid vesicle disturbance. A similar vesicle-to-bilayer transition can also be observed at such a high PEG concentration at $m = 14$ and 16.

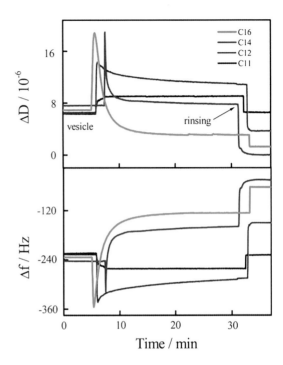

Fig. 5.8 Shifts in frequency (Δf) and dissipation (ΔD) as a function of time for the adsorption of PEG chains onto the layer formed by lipid vesicles, where the overtone number n = 3 and the polymer concentration is fixed at 2.0 mg/mL. PEG with m carbons in the hydrocarbon end group is designated as C_m. Reprinted with the permission from Ref. [15]. Copyright 2010 American Chemical Society

5.4 Conclusion

The adsorption of hydrophobically end-capped PEG chains on the phospholipid membranes has been investigated by using QCM-D in real time. Lipid vesicles can form the s-SLB on the SiO_2 surface but retain intact on the gold surface. PEG chains with weakly hydrophobic end groups only weakly interact with either s-SLB or vesicle membrane, whereas PEG chains with strongly hydrophobic end groups can insert in the lipid membranes. The hydrophobic interaction between the PEG chains and the lipid membranes increases with the length of the hydrocarbon end group of PEG. For the intact vesicles on the gold surface, the adsorption of PEG chains with strongly hydrophobic end groups leads to a vesicle-to-bilayer transition due to the insertion of polymer chains in the lipid membrane. A higher PEG concentration is more favorable for the insertion of PEG chains in the lipid membranes.

References

1. Dorairaj S, Allen TW (2007) On the thermodynamic stability of a charged arginine side chain in a transmembrane helix. Proc Natl Acad Sci 104:4943–4948
2. Im W, Brooks CL (2005) Interfacial folding and membrane insertion of designed peptides studied by molecular dynamics simulations. Proc Natl Acad Sci 102:6771–6776

References

3. Sun JJ, Vernier G, Wigelsworth DJ, Collier RJ (2007) Insertion of anthrax protective antigen into liposomal membranes-Effects of a receptor. J Biol Chem 282:1059–1065
4. Alberts B, Bray D, Lewis J, Raff M, Roberts K, Watson JD (1989) Molecular biology of the cell. Garland, New York
5. Engelman DM (1996) Crossing the hydrophobic barrier: Insertion of membrane proteins. Science 274:1850–1851
6. Shillcock JC, Lipowsky R (2005) Tension-induced fusion of bilayer membranes and vesicles. Nat Mater 4:225–228
7. Lasic DD, Needham D (1995) The "Stealth" liposome: A prototypical biomaterial. Chem Rev 95:2601–2628
8. Zhang LF, Granick S (2005) Slaved diffusion in phospholipid bilayers. Proc Natl Acad Sci 102:9118–9121
9. Lee SM, Chen H, Dettmer CM, O'Halloran TV, Nguyen ST (2007) Polymer-caged lipsomes: A pH-responsive delivery system with high stability. J Am Chem Soc 129:15096–15097
10. Venturoli M, Smit B, Sperotto MM (2005) Simulation studies of protein-induced bilayer deformations, and lipid-induced protein tilting, on a mesoscopic model for lipid bilayers with embedded proteins. Biophys J 88:1778–1798
11. Harzer U, Bechinger B (2000) Alignment of lysine-anchored membrane peptides under conditions of hydrophobic mismatch: A CD, N-15 and P-31 solid-state NMR spectroscopy investigation. Biochemistry-us 39:13106–13114
12. Lee AG (2003) Lipid-protein interactions in biological membranes: a structural perspective. Bba-biomembranes 1612:1–40
13. Keller CA, Kasemo B (1998) Surface specific kinetics of lipid vesicle adsorption measured with a quartz crystal microbalance. Biophys J 75:1397–1402
14. Keller CA, Glasmastar K, Zhdanov VP, Kasemo B (2000) Formation of supported membranes from vesicles. Phys Rev Lett 84:5443–5446
15. Zhao F, Cheng XX, Liu GM, Zhang GZ (2010) Interaction of hydrophobically end-capped poly(ethylene glycol) with phospholipid vesicles: The hydrocarbon end-chain length dependence. J Phys Chem B 114:1271–1276
16. Liu GM, Fu L, Zhang GZ (2009) Role of hydrophobic interactions in the adsorption of poly(ethylene glycol) chains on phospholipid membranes investigated with a quartz crystal microbalance. J Phys Chem B 113:3365–3369
17. Hait SK, Moulik SP (2001) Determination of critical micelle concentration (CMC) of nonionic surfactants by donor-acceptor interaction with iodine and correlation of CMC with hydrophile-lipophile balance and other parameters of the surfactants. J Surfactants Deterg 4:303–309
18. Liu GM, Zhang GZ (2005) Collapse and swelling of thermally sensitive Poly(N-isopropylacrylamide) brushes monitored with a quartz crystal microbalance. J Phys Chem B 109:743–747
19. Seifert U, Lipowsky R (1990) Adhesion of Vesicles. Phys Rev A 42:4768–4771
20. Lipowsky R, Seifert U (1991) Adhesion of Vesicles and Membranes. Mol Cryst Liq Cryst 202:17–25
21. Seifert U (1997) Configurations of fluid membranes and vesicles. Adv Phys 46:13–137
22. Reviakine I, Brisson A (2000) Formation of supported phospholipid bilayers from unilamellar vesicles investigated by atomic force microscopy. Langmuir 16:1806–1815
23. Thid D, Benkoski JJ, Svedhem S, Kasemo B, Gold J (2007) DHA-Induced changes of supported lipid membrane morphology. Langmuir 23:5878–5881

Printed by Publishers' Graphics LLC
LMO130913.15.14.59